The Harnessing of Power

The Harnessing of Power:

How 19th Century Transport Innovators Transformed the Way the World Operates

By

Maxwell Gordon Lay

Cambridge Scholars Publishing

The Harnessing of Power: How 19th Century Transport Innovators Transformed the Way the World Operates

By Maxwell Gordon Lay

This book first published 2018

Cambridge Scholars Publishing

Lady Stephenson Library, Newcastle upon Tyne, NE6 2PA, UK

British Library Cataloguing in Publication Data
A catalogue record for this book is available from the British Library

ISBN (10): 1-5275-1095-6
ISBN (13): 978-1-5275-1095-1

This book is dedicated to my dear wife Margaret who has supported me throughout its creation and production, to Robbie and Jasmine who provided constant reality checks, to all the librarians who have aided me at all times and to the wonderful access facilities provided to all modern authors by the services of Google, Wikipedia and Amazon.

A chariot can be made that moves with incredible speed without horses.
Roger Bacon, c1270

CONTENTS

Part 4: The 19th Century British Road System Responds

Part 5: Making and Using Iron

Part 6: The Black Country

Part 7: Technical Education

Part 8: Steam: A New Source of Useful Power

Part 9: The Transport Revolution is Underway

Part 10: The First Commercial Railway

Part 11: The Consequences of the New Railways

Part 19: The Context of Change

PREFACE AND INTRODUCTION

I was led to write this book as I prepared and presented an international review paper on the factors that had influenced road development in the 20th century.[1] I drew somewhat glibly on the 19th century legacy that provided the basis for most of those 20th century events. The question that increasingly came to my mind was how that creative 19th century legacy had arisen? How did transport reach the transformative stage that it did at the end of the 19th century – so many changes, so many new technologies, so many new markets?

As I thought more about these matters I came to realise the astonishing extent of the transport changes that had occurred in the 19th century. Nothing like it had ever occurred before, and the 20th century was mere incrementalism compared with the quantum changes of the 19th century. What a story there might be to uncover and to use to shed more light on our transport inheritance – where did all the 19th century's key transport features come from and why are there so many inventions, innovations, inconsistencies and illogicalities in the story?

Hence this book, which is my attempt to understand the origins of much of our current transport world. I write it as an engineer specialising in transport and certainly not as an historian additionally steeped in matters of economics, or politics or social structures. There were many great and fundamental changes occurring during the 19th century and my transport specialisation was just one sub-set of those changes. It is appropriate therefore for me to use the words of a leading English historian – Kitson Clark – to provide a broader context for my story. He described events of the 19th century as:[2]

> *"a larger movement in history ... which went on throughout the 19th century ... and which swept through human affairs and carried away the ancient régime with its aristocracies, its hereditary monarchies, its prescriptive rights and left in its stead a world whose values, on the whole, we still accept."*

I have written this book in my home town of Melbourne, Australia. Melbourne was not founded until the mid-1830s so I can view the 19th century without a strong parochial view. Indeed, my self-assigned task was to take world view of 19th century transport changes. Nevertheless, much of the book has a strong British orientation as without question the Industrial Revolution was a period and a process predominantly initiated and implemented in Britain. However, when the Revolution lost momentum in mid-century – or entered its second stage in the view of some commentators – Britain steadily began to lose its leadership role. By the end of the century France and south-western Germany were dominant change-makers and the USA was appearing on the horizon. How Britain lost, or even threw away its leadership, is a story that I explore within the wider thrust of this book.

Finally, I have tried wherever possible to highlight and pay homage to the many individual inventors and innovators and entrepreneurs who caused the dramatic transport changes that occurred during the 19th century. It will become apparent that they did this through individual initiatives to satisfy personal rather than corporate or national goals. And they were often hindered rather than aided by the various arms of officialdom.

I should note that after I began writing this book I came across a remarkable book published in 2005 by Smil covering a similar topic but with different emphases for the period from 1870 to 1914. I see this work as complementary rather than contradictory.

Note on units used. The text employs the ISO SI (Systéme International) metric units used in most countries but not, unfortunately in Britain and the USA. The main different unit is the kilometre, so to obtain miles or miles per hour, multiply the number by 0.6. For mass, I have assumed that British ton and the SI tonne are identical, whereas there is actually a 10% difference. For engines I use the SI Hertz which is a cycle per second and therefore equivalent to 60 RPM as RPM uses minutes rather than seconds. For power, the SI unit is the Watt and to convert kW to horsepower, multiply by 1.3.

Endnotes

[1] Lay 2007

[2] Kitson Clark 1962:33

PART 1

THE 19TH CENTURY'S TRANSPORT INHERITANCE

CHAPTER 1A

A LOW SPEED WORLD

The word *transport* embraces the processes that occur when people and objects move from place to place in a purposeful way. Even the earliest human communities required transport to collect and distribute food and water. However, transport is rarely a simple task and mankind has frequently sought ways to modify and improve it.

One noticeable early transport change occurred when our ancestors began to travel across the ground by walking on two legs. Once this major change had occurred, transport then changed at a minimal rate over the millions of years between that event and the beginning of the nineteenth century. This book explores the minimal changes that did occur and then focusses on the causes and consequences of the dramatic transport-related changes that arose during the 19th century. The focus on transport will not produce an introverted world view for it has been well said that "*the vital significance of improved transport to economic development is one of the few general truths which it is possible to derive from economic history.*"[1]

Whilst two-legged walking and running delivered many benefits, they had obvious transport short-comings. If we compare our unaided human transport performance when travelling on land with our equivalent performance when riding a simple bicycle (Chapter 2) or a skate-board, it is clear that as transport devices humans soon reached an evolutionary dead-end. The mechanical process involved in two-legged walking and running are inherently inefficient as, at each step, the human body mass is moved with much wasted effort in directions other than the travel direction.[2] Further, compared with the simplest hinge, the joints in our human limbs are poor pieces of mechanical engineering. Even if such inefficiencies were removed, there would still be no prospect of humans matching the minimal energy needs of a rolling wheel or of a sliding sled.

Humans walk and run at about the same speed as many other land-based animals. In terms of energy consumption, the optimum travel speed for a human is about 6 km/h, which most people would consider to be a fast walk.[3] Adults change from a walking to a running gait at around 9 km/h.[4] Trained runners can initially move at about 40 km/h but after 300 m most runners drop to an aerobic balance speed closer to 25 km/h.[5] The current world record for a marathon implies an average speed of about 20 km/h over two hours. Smil[6] gives a comprehensive review of human speed, energy and power capabilities.

A few animals can travel over land faster than humans for short periods but, as travel times lengthen, the number of animals that can out-distance the human quickly diminishes. Indeed, over many days the human can out-distance all other land-based animals. A common rule of thumb was that a *footman* could run down a horse in several days and the best horse in about six days.[7] In the 18th century a single skilled footman could travel about 120 km per day. For example, in the 1780s the Compte de Ségura used a footman to exchange a treaty between Russia and France. The courier covered the 1080 km from St Petersburg to Paris in 18 days, or 120 km per day.[8] This averaged at walking speed over the entire 24 hour day. Indeed, in most pre-19th century cases personal travel occurred at walking pace.

In pre-19th century Europe most journeys from town to town for most people would take at least a day. The major exception was in the large migratory movements that occurred with seasonal variations in climate. Whilst these could involve large distances, they nevertheless still occurred at walking speed.

The ability to routinely travel long distances quickly was not an important human need and thus received little evolutionary priority as we evolved in what we would now consider to be a low-speed world. One associated consequence is that we have a poor visual ability to determine the speed of an oncoming object and we then take many seconds to react to our observation. Instead, we have evolved the ability to detect and identify another creature a kilometre away, giving us minutes to decide what to do before an encounter might occur. Likewise, we have not needed to evolve any significant resistance to impact by large moving objects. In our low-speed world the threats and opportunities have been relatively local and so we react quickly to nearby unexpected events and have a remarkable ability – called *peripheral vision* – to detect moving objects at right angles to the direction in which we are looking. The various other human factors that affect

In the low-speed world it was widely believed that ill would befall any humans who travelled at higher speeds. For instance, it was commonly believed that air breathed in when travelling at speed would crush the lungs and the brain would be forced into permanent hallucinations. People who fell off cliffs would never experience any pain on final impact as they would already be dead as the speed of the fall would have made it impossible for them to breath. The view that rapid motion could cause apoplexy was behind an 1837 Whitehall instruction to a judge travelling by horse-drawn coach from London to Edinburgh that he should at least stop for the night at York for "*if you go all the way by that coach it will force the blood into your head and you will die*".[9] In 1840 a then well-known British writer on science and technology, Dionysius Lardner, claimed that people travelling at speed of 190 km/h (in an out-of-control steam locomotive) would be unable to breathe and would therefore die from suffocation.[10]

A lack of speed perception did cause many fatalities and serious injuries at the beginning of the steam train era (Chapter 12). For example, the case of a major British politician being very publicly run-over and killed when he misjudged the speed of an oncoming train is discussed in Chapter 10b. Similar problems arose when people leaned out of a moving train and were struck by a trackside object or when they stepped from a train before it had completely stopped. Dionysius Lardner had properly advised early train travellers "*Never attempt to get into or out of a railway carriage when it is moving, no matter how slowly, for it is a peculiarity of railway locomotion, that the speed, when not very rapid, always appears to the unpractised passenger to be much less than it actually is*."[11] The advice remains valid today and, as can be seen with young children, judging speed remains a learnt rather than an inherited evolutionary ability.

Transport often involves moving objects. Humans struggle to carry loads which are only a fraction of their body weight. Other "stronger" creatures that can carry loads without external help can rarely move more than half their body weight.[12] By comparison, the load capacity of a wheeled platform can be many, many times greater than the weight of the platform. The next Chapter will describe how, as agriculture developed, this human inadequacy led to the first transport invention and the need for humans to carry heavy weights continued to receive little evolutionary priority.

Clearly, improvements in transport, specifically in travel speed, in load-carrying capability, and as a tool for war and conquest, required invention rather than evolution. Thus transport invention was driven largely by the needs of increasingly organised communities to move freight for food and trade, to meet the demands of warfare and to efficiently administer large territories. It is appropriate to now briefly describe that process as it continued over the millennia, before peaking in the 19th century – a period which will be described in much more detail in later Chapters.

CHAPTER 1B

THE FIRST TRANSPORT INVENTIONS

Our evolved personal transport capabilities were unable to meet the needs of developing communities. Consequently, transport improvements started to occur through invention rather than evolution. The first major improvement followed from the discovery in about 5000 BC that some breeds of cattle could be domesticated and that castration made it possible to train them to do useful tasks. This particularly applied to oxen which had prominent shoulders which made it relatively simple to develop harnesses that enabled the animals to pull the primitive ploughs needed to extend agriculture beyond a subsistence level. It was then a small step to move from pulling a plough through soil to dragging a sled able to transport useful loads over smooth ground.

Oxen were strong and able to operate on poor ground, however they moved very slowly at less than human walking speed. Another six millennia were to pass before horses could be used for haulage as a horse does not have prominent shoulders and its harness must be carefully designed, constructed and operated to avoid pressing on its windpipe. Many societies were unable to do this, and effective horse harnesses for freight applications were not widespread until about 750 AD.[13] Horses were also extensively used to carry freight on their backs and could manage loads of about a third of their body weight.

Moving an object requires overcoming *forces* due to friction and gravitation. Technically, the *energy* needed is the product of these forces and the travel distance. The effort needed to haul an object is indicated by the time rate at which the haulage task consumes this energy and is called motive *power* and is fundamental to the story in this book. Indeed, it has been argued that power as we know it is the great discovery of the 19th century.[14] It is measured in Watts (or Joule/second) and has the symbol W. Humans can produce about 50 W whereas oxen and horses provide about 500 W. This order of magnitude improvement highlights the importance of learning how to harness animals and utilise their power. Indeed, the first practical internal-combustion engines still only provided about 500 W (Chapter 14). Although an ox could, typically, haul a larger load and thus technically do more work than a horse, the greater speed of the horse meant that it usually provided more power than the ox.[15]

A sled dragged slowly across ice is a situation requiring minimal motive power. There are obvious limits to the usefulness of this application and the frictional forces produced when dragging a sled across land can be very high. The alternative is the wheel which avoids the large friction forces by ensuring that most relative movement occurs in the controlled environment between the wheel hub and the fixed axle. By 3000 BC a number of wheeled devices had been invented and operated.

A few definitions to be used in this book should be introduced here. A cart is a two-wheeled and a wagon is a four-wheeled load-carrying vehicle. A dray is a cart without sides. In the 19th century the more common spelling was "*waggon*". A carriage is a vehicle for conveying people and was called a coach if the passenger space was enclosed. These three vehicle types do not have their own power source. A car is a self-powered carriage. A truck is a self-powered wagon. In Britain, the word "*lorry*" is sometimes used instead of "*truck*". The text uses the British term tyre rather than "*tire*", although "*tire*" was the original 19th century word and has a correct linguistic basis. "*Tyre*" became common in Britain in the 1930s.

The wheel was not a simple development and wheeled vehicles never occurred in a number of otherwise advanced communities. Furthermore, the technology for making or the wealth for buying wheels and wagons was not widespread. In 1800 at the start of our revue period, there were still parts of Scotland, Wales and the English west country where wheeled vehicles were unknown.[16]

When the technical challenges were met, the wheeled vehicle brought many advantages and improvements. However, for the five millennia following its first invention, the wheel provided a far from perfect outcome. Wheels were difficult to make, they were often cumbersome, they separated from their axles, and their axle hubs and running surfaces wore very quickly. In the 1790s cast iron (Chapter 5a) was used to provide oil-lubricated joints between the rotating wheels and their fixed axles, thus dramatically reducing a millennia-old operating problem. The method was patented by John Besant in 1795 and was first used on Britain's Royal Mail coaches as will be described in Chapter 1e.

There were two further, related, problems. The stiffness of the wooden or iron running surface of the wheels meant that very high contact pressures occurred between each wheel and any hard and trafficable road surface, causing both to wear rapidly. On the other hand, if the road surface was relatively soft, the wheels would sink into the roadway, greatly increasing the traction forces needed and making the road far less trafficable.

These contact pressure problems could be alleviated by making the wheel running surface wider, but the wheel then became much heavier. Another common solution was to make the wheel radius bigger which increased the effective contact area between the wheel and the road and thus reduced the damaging pressures. However it also raised the body of the vehicle making it difficult to load. It also made it impossible to have a steerable front axle for the wheel could not fit under the vehicle when the steering axle turned about its central vertical axis (Fig. 1).

Fig. 1 A 19th century wagon with high wheels, demonstrating their effect on wagon height and steerability. From Lane[17].

The vertical loads produced by the mass of the wagon and its freight bore on the pavements via the wagon wheels and often resulted in significant pavement damage. This led to much argument and a great deal of empirical and fruitless pursuit of dead-end developments, such as the contorted production of dished and conical wheels to carry heavy loads.[18] A variety of wheel shapes had been proposed and used during the 18th century (Fig. 2) and the debate about the appropriate system was still raging at the end of the 19th century. The issue was finally swept away by the wide-spread use of the pneumatic tyre early in the 20th century (Chapter 16a). These tyres automatically ensured that contact pressures did not exceed the inflation pressure of the tyre. Nevertheless, in the 19th century most motor trucks ran on wheels with solid rubber running surfaces, which were only slightly better than iron surfaces.

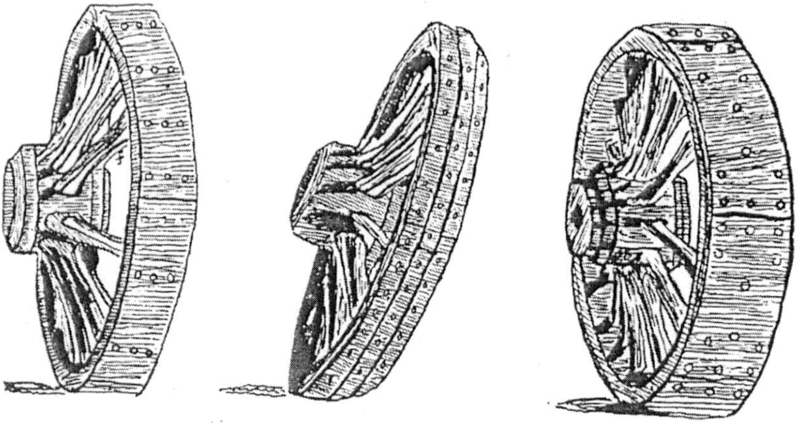

Fig. 2 Wide dished and conical wheels developed prior to the pneumatic tyre in order to minimise contact pressures and to carry heavy loads. See Fig. 4 for a further explanation of the wheel in the centre. From Pyne[19].

A wagon weighed about a ton (about the same as a modern car) and could carry up to 8 ton.[20] A typical loaded wagon would travel at walking speeds or less and its movement was truly "*ponderous*". The great wagon journeys across the USA in the 1840s averaged about 25 km per day or about half walking speed.[21] Whilst ascending even a gentle up-slope was a problem for the heavy vehicles, descending hills was often more of a problem, as the lack of adequate braking meant that vehicles could easily begin rolling out of control towards the hauling beasts or into roadside ditches, with disastrous consequences.

Wheels were not essential for improved personal land transport. By 3000 BC horses had been tamed sufficiently to permit horse riding to occur. This was followed by the use of light two-wheeled chariots. Selective horse breeding was widely practised, and specific lines were bred for a range of purposes.

The use of reins attached to transverse pieces (*bits*) in the horse's mouth greatly increased the control that a rider could exert over a horse. Metal "*bits*" were not invented until about 900 BC. The introduction of foot stirrups in about 200 BC finally made horse riding an easy means of personal transport and greatly improved the weapon-wielding effectiveness of a warrior on horse-back. With good reins and stirrups and a strong horse, a single horse and rider could then manage about 70 km per day. In conjunction with a well-organised stage system using a series of horses, messages and small packages could be carried about 250 km per day, at an average speed of around 10 km/h. This was about twice the speed of a single man on foot, but did require pre-organised stages. Furthermore, stabling was required along the route as at each staging stop the rider's horse was replaced by a fresh, rested horse.

CHAPTER 1C

ALTERNATIVES TO THE WAGON

As societies moved beyond subsistence levels, the need to move products led to a steady increase in the use of animal-drawn wagons for freight haulage. However, at the best a horse could haul a load of about a ton over short distances or half a ton over long distances. Consequently, freight capacities were extremely small and the pressure for change came more from a need to increase the size of the loads that could be carried than from a need to increase travel speeds.

The movement of freight is of paramount importance to any developing economy. It has been estimated that the haulage task on roads servicing London grew by a factor of over three between 1800 and 1840.[22] Textiles were a major contributor to this increase. It is relevant to underline that this rapid growth was occurring before the railway era began. At the advent of railways in England, there were over 2000 regular wagon services per day.[23] Assuming a modest load of 5 ton (Chapter 1b) suggests that at least 10 000 ton a day was being moved on Britain's pre-railway roads. Whilst much of this growth was due to the new industrial processes, some was also a consequence of the end of the Napoleonic Wars in 1815 and the subsequent refocussing of attention on internal developments.

Significant amounts of freight were also moved by water. There is a long history of using sailing ships and galleys to move freight between Mediterranean ports.[24] In Britain there were regular movements of grain and coal along the east coast. The tonnage capacity of the ships engaged in British coastal shipping increased from 330 thousand ton in 1790, to 830 thousand ton in 1824. This is much faster than the general growth of economic activity during the same period. In 1824, some 5 million ton of coal and half a million ton of grain were moved by British coastal shipping.[25]

Although these very large volumes and tonnages of freight could be moved by sea between suitable ports, the process could also be extremely slow, as the basically square-rigged sailing ships predominantly travelled down-wind and were thus exposed to the vagaries of the weather and as – without adequate time-keeping – navigation was often haphazard and wildly inaccurate.[26] Ships rarely exceeded walking speed and even in the 18th century ships sailing from England to Australia in favourable conditions typically had average speeds of about 10 km/h.

In some inland locations, large quantities of freight could be floated down rivers and since the Middle Ages many European rivers had been modified to make this process more effective. Large communities had developed canal-like systems for water supply and drainage. In order to enhance the economic potential of tracts of countryside, major inland canals capable of carrying horse-drawn barges were built in Britain and continental Europe from the mid-16th century onwards to a usage peak in the 18th century.[27]

Many of the British canals were built by groups of speculators and investors and their main customers were the producers and consumers of coal. For example the Bridgewater Canal (Chapter 10a) had been built in the 1760s by James Brindley to take coal from the Duke of Bridgewater's mines at Worsley to Manchester, some 10 km to the east. It crossed the River Irwell on an aqueduct. Operation of the canal halved the price of coal in Manchester. Within a few years there were major extensions east to Runcorn on the Mersey River estuary and to collieries at Leigh. The horse-drawn canal system that resulted from such investments provided the transport backbone for the initial phases of the Industrial Revolution.

Large-scale canal development also occurred in the USA between the 1812-4 war against Britain and the beginning of the railway era a few decades later. Major works such as the Erie and Pennsylvania canals were built to enable America to peacefully develop its vast lands east of the Appalachians.

Typically, one canal barge carried as much freight as 12 horse-drawn wagons. Haulage capabilities demonstrate the difference between transport modes. At the beginning of the 19th century a well-harnessed horse could pull a load of 1 ton on a good level road for up to 5 km. For longer distances, the capability dropped to about 0.5 t. When iron rails became available (Chapter 9c), the smoother surface meant that the same horse could then pull a load of 8 t. Further improvements in haulage capacity occurred with canals and their tow paths, where the flat gradient and low resistance to motion allowed a horse to haul 50 t loads.[28] An important factor with canals is that the resistance to motion depends only on the cross-sectional area of the barge, and so the resistance per ton carried reduces as the load hauled and the barge length increase (Chapter 9f).

In pre-railway England, many of canals were very profitable investments. For example, in 1825 the ten best performers returned an average annual dividend of 28%. A major cargo was coal produced at

England's inland coalfields and needed throughout the country, particularly as the use of coal gas for power generation and lighting became more commonplace (Chapter 15a). Canals also provided passenger services for, although much slower than wagons, they were far more comfortable, and in 1836 the Union Canal was carrying about 600 passengers per day between Edinburgh and Glasgow.[29] Passenger services were also improved with the introduction of long and light shallow draft boats. When hauled by trotting horses these "*swift boats*" could reach speeds of 15 km/h, surfing on their own bow waves.

The above figures neatly illustrate a prime reason for investing in transport facilities, even in the absence of mechanical power. That reason is that maintaining momentum is a relatively undemanding task when the resisting forces of friction and gravity have been minimised by appropriate transport infrastructure.

CHAPTER 1D

PERSONAL TRAVEL

For most people, walking was the only available means of travelling within towns and between towns. It was still a very common means of long-distance travel in the 18th century. For instance in England many soldiers would be encountered being redeployed around the country or travelling to and from training camps and embarkation ports.

A revealing mid-18th century travel example occurred when the famous blind English road-builder, John Metcalf (Chapter 4a), was in London with his local member of Parliament, a Colonel Liddle. Liddle offered Metcalf a ride home to Knaresborough in his carriage, a journey of some 300 km. Metcalf declined and said he preferred to walk home. Blind, on foot, and unfamiliar with route, he completed the journey in 6 days. The Colonel and his carriage took 8 days.[30]

Personal travel by riding on the backs of animals or in animal-drawn vehicles was a matter of perceived convenience and not of time saving. Until the 18th century, there was little difference between vehicular personal travel and freight haulage as both used wagon-style vehicles which travelled at walking speed (Fig. 3). Much of the technology appears to have been of Celtic origin. The Celts began building light and sophisticated wheels in about 700 BC.[31] Key features were spoked wheels with applied wear-resistant running surfaces. Iron was widely used in Celtic vehicles and harnesses. A prime example of an early Celtic vehicle with many advanced features is the Dejbjerg wagon from about 50 BC and found well-preserved in the 1870s in a Danish swamp (Fig. 3a). The reconstructed Roman passenger vehicle in Fig. 3b is quite advanced as it also has strap suspension and a swivelling, steerable front axle.

Fig. 3a A reconstructed Celtic vehicle from about 50 BC. It was found in a Danish swamp at Dejbjerg in the 1870s. In the Danish National Museum. Photo, wordpress.com via Wikipedia Commons.

Fig. 3b A reconstructed Roman passenger vehicle in Cologne Museum. Both the front axle and the harnessing shaft pivot around a central vertical pin. Photo with kind permission of the Rheinische Bildarchiv, Cologne.

Wagons for transporting people came into prominence in the 13th century although – as shown in Fig. 3 – many of the innovations were reinventions of much earlier Celtic technology. A key need was to provide passengers with a smooth ride. However, a wagon with four rigidly attached wheels will always provide a rough ride on road surfaces which are not perfectly flat planes. This is because, as a matter of geometry, the rigid wagon can only make contact at three points. The fourth point is poised above the road surface, ready to descend with a rattling bump. The Celts found a way around this problem. They had developed good leather-working skills for harness making, and were able to make leather straps to suspend the vehicle cabin, as can be seen in Fig. 3b. This reduced the previous abruptness of the bumps and the Roman philosopher Seneca said that his carriage travelled so smoothly that he could write whilst on a journey. The term "*suspension*" is still used to describe ride-softening devices in modern vehicles, although they no longer work as suspension devices.[32]

Vehicle technology only slowly improved over the next millennium. The main developments were better harnesses and lighter and stronger wheels. Increases in travel speed above walking began to occur early in the 17th century, mainly as a consequence of greater demands by the wealthy for personal travel and, at the supply end, to a wider use of iron in vehicle construction. This use of iron introduces a main thesis of this book (Chapter 6).

Iron had many useful transport applications. However, iron making was still a craft rather than a manufacturing process, so its spreading use in vehicles indicated strong market demands. As discussed in Chapter 1b, the use of iron for bits and stirrups greatly enhanced the usefulness of horse riding. For vehicles, iron rivets simplified the attachment of the hauling part of the harness to the vehicle being hauled. Iron studs and plates were used to prevent the rapid wear of wooden rims and iron pins were used to retain the wheel on the axle. The wearing surface between the wheel and the axle was a major operating problem. The bearing surfaces themselves were heavily lubricated with animal fats and occasionally employed some cylindrical metal pieces although these were more likely to have been the softer bronze, rather than iron.

Iron coil springs were invented in England in 1625 and were soon used to support coach bodies at each of their four corners. A major advance demonstrating the increasing availability of useful pieces of iron occurred in the 1660s in Prussian Berlin where Philip di Chiesa, a person of Italian extraction born near Avignon, greatly improved the rideability of coaches by hanging the leather suspension straps from cantilevers made of flat iron bars (Fig. 4).[33] Such leaf springs are still used in many current vehicles. Di Chiesa's vehicles were called *Berliners*. They were relatively large with an enclosed passenger area. The leather straps were subsequently replaced by iron links.

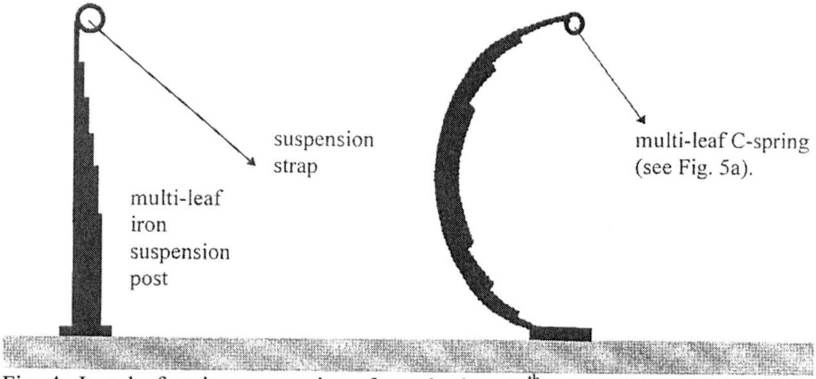

Fig. 4 Iron leaf spring suspensions from the late 17th century. Depending on its shape, the C spring might also be called a circular or an elliptical spring.

Germany was then the centre of quality coach-building. Increased competition from the German *Berliners* led to further developments in Britain, including the work of the multi-talented Robert Hooke and his definition and scientific measurement of the elasticity of iron in the 1670s.[34] Elliptical leaf springs (Fig. 4) were developed by Obadiah Elliot[35] in London in 1804 and Chapter 1f describes how they produced a quantum jump in coach performance. Elliot was a well-known coach-maker who patented his invention in 1805. The introduction of iron rim brakes in 1690 allowed larger teams of horses to be used to power the coaches. However, braking as we now know it was never a serious issue at the speeds of a horse-drawn coach and the brakes' major role was to slow vehicles descending steep hills.

CHAPTER 1E

THE GOLDEN AGE OF COACHING

By 1750 coach design had improved to such an extent that the coach had taken over from horseback riding as the dominant means of long-distance personal travel. Nevertheless, travel speeds were less than 15 km/h and a day's journey would not exceed 120 km. At the end of the 18th century, English coach-builder William Fenton wrote "*the art of coach building has been in a state of gradual improvement for half a century past, and had now arrived at a very high degree of perfection, with respect to the beauty, strength and elegance of our English carriages.*"[36] Our perceptions of horse-drawn vehicular personal travel are largely coloured by modern images of the early American stage coach era, however the 18th century reality shown in Fig. 5 was somewhat different.

Fig. 5 A stage coach as drawn by Pyne[37] in 1806 or earlier. Note that the smaller front wheels are able to pass under carriage body during turning (see also subsequent Figures). The wheels usually toe in at the bottom so that the vehicle weight forces them inwards on the axle. This requires the wheels to be dished so that the lowest spoke – which carries much of the load – is vertical (see also Fig. 2).

The operation of the Royal Mail system in Britain provides a useful example of the more pragmatic changes that were also occurring. For much of the 18th century the service had used post-boys running on foot or riding horses and reaching peak speeds of about 10 km/h. In 1784, a Bath entrepreneur and theatre proprietor named John Palmer, dissatisfied with the current system, proposed a different operating model.[38] The basic points of his proposal were to use a new form of coach carrying four passengers and the mail in a special compartment. The fares paid by the passengers would be sufficient to fund the commercial operators of the service. In a major change, the coaches would operate to a defined timetable with only short stops at coaching stations. Armed guards would be used to prevent "*highway men*" from waylaying and then robbing coaches and their occupants. Whilst the coachman worked for the coach owner, the guards were government employees. The proposal gained the support of the new Prime Minister, William Pitt, and the Post Office quickly introduced the new postal system.

The new service began operating in 1785 at average service speeds of about 13 km/h, which was a significant improvement over existing services. In 1786 Palmer was appointed Controller General of the Post Office. In that year the Post Office accepted the design of a new Berliner-style coach produced and recently patented by John Besant[39], a London coach-maker. His "*Patent coach*" (Fig. 6) was much better than its predecessors as it had improved braking, continuous lubrication of the axle bearings and more secure attachment of the wheels to the axles. However, it was still had the high centre of mass that made the coaches prone to overturning. Besant's contract included the supply and maintenance of the coaches. Initially there were twenty Royal Mail services leaving London each evening and Besant soon found that this required a fleet of forty coaches. By 1792 the fleet had tripled in size.[40]

At its peak the Mail's horse-drawn services were operating at about 25 km/h – much faster than the equivalent horse-back service – and operated in the most challenging of circumstances (Fig. 7). The combination of faster mail coaches and a well-organised postal service had widespread effects. For example, it halved the time it took to have business and governmental decisions made and implemented within a regional structure. More broadly, Anderson notes that the "*coaching services gave a new unity to the nation*" as illustrated by the widespread excitement when the coaches spread the news that the 1832 Reform Act – which enhanced the electoral system – had passed through Parliament. London newspapers bringing detailed accounts of national and international news were distributed daily and coachmen and coach guards shouted out the headlines to people on the streets. Coaching also had major social impacts.[41] The guard on a coach had both a gun and a bugle (or horn) – the bugle caused great excitement as a coach approached a town and was used by many to correct their timepieces. The original and official purpose of the horn was to alert villagers that mail was being left at the coach-house.

Fig. 6 The first of the Royal Mail "patent" coaches. The armed guard on the left was a Post Office employee. From Vale[42].

Fig. 7 A Royal Mail coach travelling from London to Norwich in a severe storm. R. G. Reeve aquatint from Belloc[43].

A natural consequence of the improvements in coach performance was a major increase in the services offered, as Fig. 8 shows. However, the number of passengers carried is even greater than indicated, as changes in coach design after 1790 allowed more outside passengers and doubled the carrying capacity of the coaches (compare Figs 5 & 7). The service decline after the 1830s reflects the impact of the new railways (Chapters 11).

On the London – Manchester run in 1790 a well-known coach-master, Edward Sherman,[44] introduced his Beehive coaches which enhanced user amenity by using numbered spring-cushion seats, an interior lamp for the overnight journey and flattened leaf springs. Services continued to improve and there was soon little distinction between one of these inter-town coaches and a horse-drawn urban bus (Chapter 12b).

Concord stage coaches were developed in America early in the 19th century as sturdy and reliable vehicles able to service long-distance routes. They used strap suspension (as in the Roman wagon in Fig. 3) and this gave passengers a smoother ride than the British Royal Mail coaches. It is appropriate to add that in the preceding century the large and lumbering Conestoga wagons – or "*prairie schooners*" – had moved freight at walking pace across the vast American landscape.

Although the flattened leaf spring had allowed some lowering of the centre of mass, each of the vehicles shown in Figs 5, 6 & 7 were still quite high off the ground by today's standards. The three main reasons for this persistent design feature were:

(1) large diameter wheels were needed to reduce both the pressures between the hard surfaces of the wheels and the pavements and the necessary traction forces (Chapter 8). Extensive tests conducted by Morin[45] in France in the 1830s had indicated that the traction force was inversely proportional to the wheel diameter,[46]

(2) a mistaken belief that raising the centre of mass reduced the haulage task of the horses, and

(3) for personal-use vehicles such as those shown later in Figs 10 & 11, to make a major fashion statement.

The high position of the centre of mass created routine difficulties for passengers boarding or leaving the vehicle. Another major consequence was that there was the distinct likelihood of such a vehicle overturning, as implied in Fig. 9. The resulting crash rate was a worry to many and in 1811 and 1832 Britain passed laws restricting the laden height of vehicles to about 3.3 m. Operating crashes resulting from other causes, such as wheel and axle failure, were also commonplace (Fig. 10).

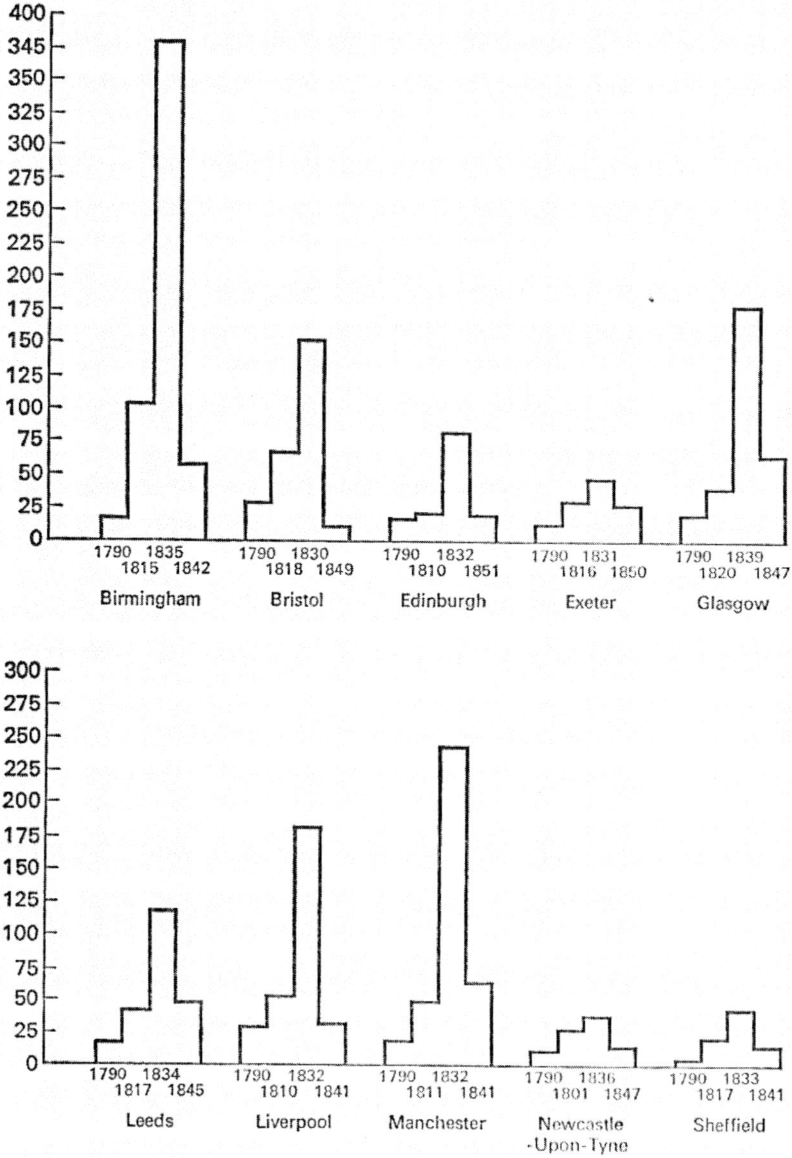

Fig. 8 British coach travel between 1790 and 1850, showing the number of daily departures of Royal Mail and stage coaches from major towns. From Bagwell[47].

Fig. 9 One of Pyne's[48] coach drawings showing a potentially perilous situation. The large luggage basket at the rear and lack of a guard indicates that it was not part of the Royal Mail system!

Travel safety did not improve until the coming of the railways (Chapter 11) when fatalities per passenger carried dropped by a factor of about 5.[49]

Increased road traffic towards the end of the 19th century once again saw road fatalities rise. For example, New York City in 1900 recorded about 100 road fatalities per million of population.[50] This is not unlike the rate experienced in modern cities today, although more of the fatalities were local pedestrians rather than travellers.

Fig. 10 A Royal Mail crash on the road to Bath. The coach is a Besant Patent. Drawn by Thomas Ronaldson (1756-1827) and titled “The night coach”. From Google Images.

The period from 1820 to 1840 is sometimes called the Golden Age of Coaching[51] and has been often described with appropriate colour and tempo, as in the following example.[52]

> “*A guard travelled on each coach armed with a blunderbuss in addition to his official horn – the music of which he preferred to supplement by playing on a bugle. It was difficult to prevent him from discharging his weapon in sheer exuberance of spirits to add to the excitement caused by the passing of his coach……The coachmen (drivers) had improved their appearance (though not always their tongues, unless mellowed by cash) and blossomed out in gorgeous waistcoats and cravats, white beaver hats, strapped trousers, the whole embellished with a fine cigar.*
>
> *….In the first quarter of the* 19th *century Britain could boast that she had the most perfect roads, the best vehicles, and the finest horses to draw them in the world. “The Road” was a great industry which provided employment for many thousands: coach proprietors, coachmen, guards, grooms, ostlers, stable boys, innkeepers, waiters, maids, bootlacks, farmers, horse breeders, coach builders, saddlers and wheelwrights*”.

Until the coming of the railways in the middle of the 19th century, stage coaches were the major form of long distance transport. In England coaching reached its peak in 1835 when there were 700 mail coaches, 3300 stage-coaches, a work-force of 35 000, and 150 000 horses. Overall, there were about a million horses involved in road transport in Britain.[53] There were 50 daily passenger coaches servicing the popular route between London and Brighton. Not coincidentally, this peak aligns with growing power of the new steam railways (Chapter 12). Thus, although the new railways did create new transport demands, they were also able to take advantage of a strong and growing set of established demands for personal and freight transport.

Some coach owners fought against the railways in a vain attempt to preserve their businesses. Others were more perceptive. A horse requires about 2 hectare of grassed land for its support, and most working vehicles needed between 10 and 30 horses to sustain one day's operations. In addition, further land was needed for horse breeding, as the working life of a horse was only about 4 year. Thus each vehicle needed about 40 hectare of good land and so operating horse-drawn services in competition with the new railways required some commercial skill.

William Chaplin[54] was the largest coach operator in England in 1835, with nearly two thousand horses and 70 coaches in his system, and was colloquially known as the "*Napoleon of coaching*". However, he was quick to negotiate with the railways to offer feeder services to their stations and then later became chairman of the London and South Western railway. For Britain, it has been claimed that "*Railways created more short-distance horse-drawn transport on roads to railway stations and in areas still unserved by trains than they eliminated over medium and longer distances*".[55] This is supported by the following data on the number of horses not on farms and being used for travel, sport and haulage, despite the considerable expense of maintaining a horse:[56]

1811	487,000
1851	535,000
1891	1,360,000
1901	1,770,000

Many of these horses were operated from stables and farms owned and operated by the railway companies.

The advent of the railways did dramatically reduce the demand for more and better coaches and the fact that, per ton-km, horse-drawn transport was about three times more expensive than railway transport, meant that there were limited opportunities for long-distance horse-drawn freight transport. Thus the further pursuit of coach and wagon development adds little to the thrust of this book and will not be pursued further.

CHAPTER 1F

TRAVEL FOR THE WEALTHY

The style of the coaches used by the wealthy changed from solid and sturdy to swift and stylish as coach and carriage design developed significantly during the late 18th and early 19th centuries. The new coaches were a major improvement over their predecessors in terms of both speed and comfort and reached their zenith in the first decades of the 19th century. The demand for change had been driven by better roads (Chapter 3), consumer preferences for more flamboyant travel, improved manufacturing methods and technical input from the embryonic learned academies in England, Germany, France and Denmark.[57]

German coach technology had been particularly strong during the 17th and 18th centuries and set the pattern for other countries. For upper class travel over short distances the light but sturdy coach style first produced in Landau in south-western Germany in the 1740s became popular throughout the 19th century and is typified by the Barouche (or caroche) style shown in Fig. 11.

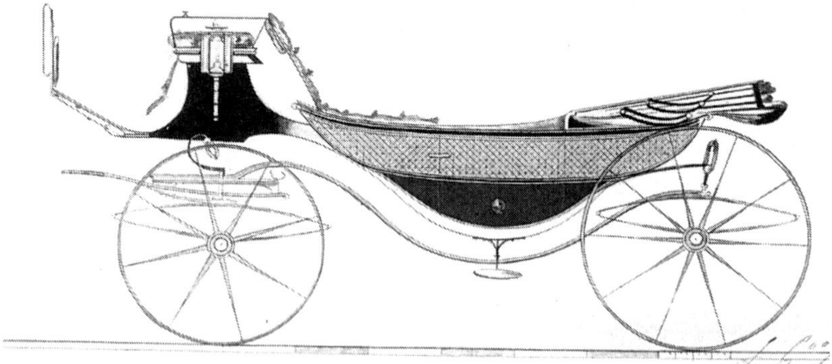

Fig. 11 A Barouche coach, c 1810. It was pulled by two to four horses, was about 3 m long with a fold-up hood and seated two to four passengers. Some versions used a strap suspension at the rear. Drawn by R Ackerman, who was a contemporary German lithographer and highly-innovative coach-builder.[58] via Wikimedia.

King George III ascended to the British throne in 1760. He had a penchant for travelling as quickly as possible and his demands for better performance saw the development of a new generation of light, elegant and high-speed vehicles as epitomised by the Highflyer phaeton carriage drawn by one or two horses. Phaetons were based on the sturdier Barouche which continued to be used for formal and social occasions. The first Highflyer was produced for the King in about 1770 and had provision for two footmen to ride over the rear axle. The vehicle was also fashionable in 18th century Paris. The Highflyers were dangerous to drive and were popular with wealthy young men, including the King's eldest son (later George IV). By 1790 the Highflyers were of the form shown Fig. 12a and could be drawn by two to six horses.[59] The two-wheeled coach version was known as the Cabriolet (Fig. 12b). The Highflyers then developed into an effective vehicle for use in providing taxi-type services in large towns (Chapter 12b).

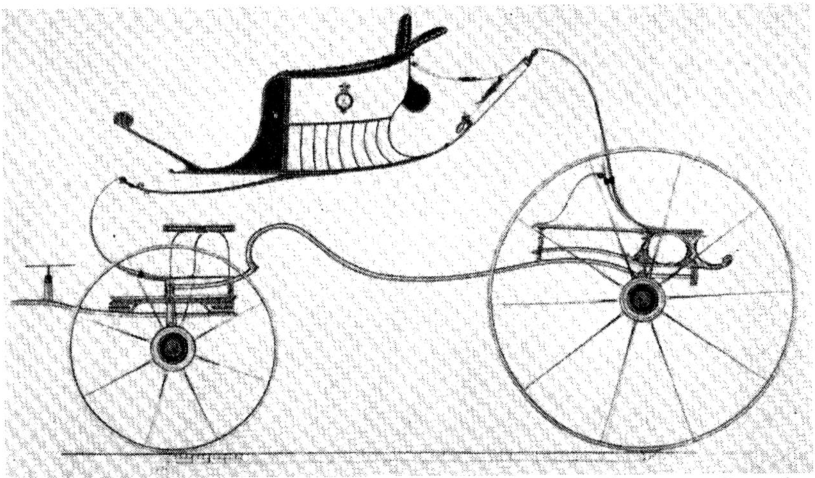

Fig. 12a The Highflyer crane-neck Phaeton, in about 1790. The name reflects the height of the seat above the ground. The diameter of the rear wheel was about 2 m. From Watney[60].

These turn-of-the-century coaches, exemplified by the Phaeton, were an outstanding example of the clever use of iron and of combined strap and spring suspensions. They were at the elegant end of the coaching spectrum and fitted appropriately within a period and place which also saw Beethoven producing some of the world's greatest music.

Fig. 12b The Cabriolet coach, in about 1800. This development from the Phaeton was a popular vehicle for use in towns. A lighter single person version was called a gig. via Wikipedia Commons.

As discussed in Chapter 1d, elliptical leaf springs (Fig. 4b) were developed by Obadiah Elliot in London in 1804 and produced a quantum jump in coach performance. They were used in pairs, either longitudinally Fig. 11) or transversely (Fig. 13) or in combination. Their key advantages were that they permitted a carriage body which was more self-supporting (monocoque) and therefore much lighter and that they lowered the centre of mass of the vehicle thus making passenger access easier and decreasing the risk of roll-over.[61]

However, following the advent of the railways, one coaching change was that there was no longer a need to carry large amounts of luggage. Thus even mundane coaches became sleeker and more sporting than their predecessors.[62] A prime example occurred in 1838 at the beginning of the railway age when Lord Brougham – the Lord Chancellor of Britain – required a smart four-wheeled carriage that could be pulled by one horse and used by "gentlemen" for journeys around town. The resulting stylish Brougham (Fig. 14) became the most popular personal vehicle in the period preceding the car. The Landau coach was a similar vehicle which relied solely on elliptical leaf spring suspension.

Fig. 13 A Goddard buggy, designed by Thomas Goddard in the USA in the 1880s. Note the transverse leaf springs. via Wikipedia Commons.

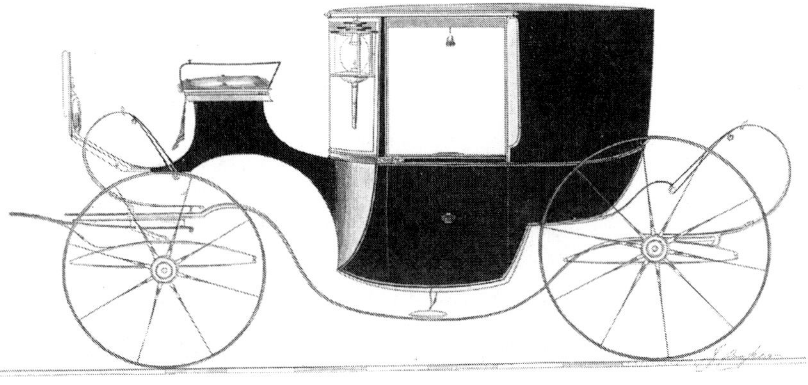

Fig. 14 J. Cooper's contemporary (c1850) drawing of a (one-horse) Brougham town vehicle. Note the lowering of the body to improve passenger access. via Wikipedia Commons.

PART 2

PEOPLE POWER

CHAPTER 2

PEOPLE POWER

Travelling at speeds faster than, and carrying loads heavier than, our evolutionary inheritance required the invention of devices to overcome these inherited transport deficiencies. The availability of the light frames and wheels that had made it possible to produce coaches such as the Barouche, Phaeton and Cabriolet shown in Figs 11 & 12 also led inventors to think of vehicles powered by humans rather than by horses and oxen. In 1777 Carl Theodor in Germany produced a gentleman's coach powered by a servant standing over the rear axle and "walking" on lever-like pedals. It was the first of a number of similar and increasingly more ingenious human-powered vehicles.

A more democratic device than Theodor's servant-powered coach was introduced in 1817 in Mannheim in Germany by Karl Drais,[63] a prolific inventor.[64] In the manner of a modern bicycle, Drais used a frame to connect two wheels in line. The machine was propelled along the road by the rider running on the pavement whilst straddling the frame (Fig. 15) and then coasting with legs raised when there was adequate momentum. It thus overcame two of the core technical inefficiencies inherent in walking and running – keeping the mass moving in the travel direction and having a low resistance to motion. The Drais device was steerable, whereas an earlier version built by Count de Sivrac in France in 1791 had lacked steering and proved impractical.

Fig. 15 Karl Drais' original vehicle for personal transport by Ginzroth[65] in 1817.

The devices were called "*laufmaschines" (running machines)*; in England they were the original "*hobby horses*",[66] in America they were "*dandy horses*.[67] They proved very popular for personal use and as local delivery vehicles (Fig. 16). Speeds of up to 15 km/h were achieved.

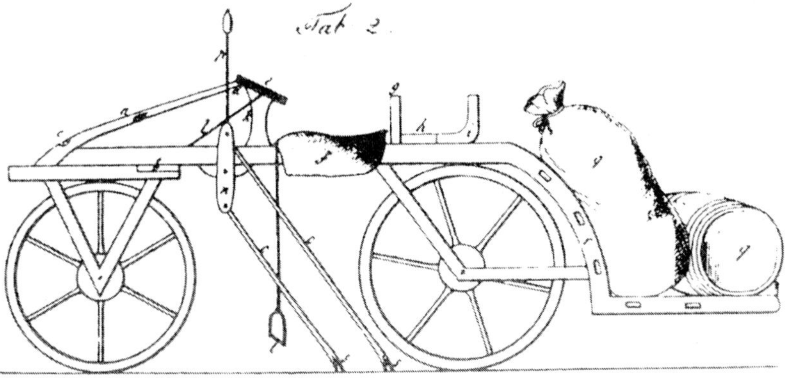

Fig. 16 Original Karl Drais delivery vehicle From Bauer[68] in 1817.

From a simple but clever way to use wheels, the next steps towards a practical vehicle required some decades of mechanical ingenuity, a process to be mirrored later and at a larger scale by both the steam train and the car. In 1830 the French postal service supplied postmen with a Drais-type vehicle powered, not by running on the ground, but by using the legs to turn cranked axles to produce rotary motion transmitted to the rear wheel. It could thus be considered to be the first bicycle. Kirkpatrick MacMillan,[69] a blacksmith, built a couple of treadle-powered bicycles in Courthill in Scotland in 1839.

The two-wheeled arrangement had been introduced to allow the rider to straddle the frame whilst stabilising the vehicle with his feet. However, as higher vehicle speeds were achieved and unexpected major side benefit became apparent. At even low speeds the angular momentum of the rapidly spinning wheels produced a gyroscopic effect that gave the otherwise unstable vehicle great vertical stability. Nevertheless, from 1840 to 1870 Willard Sawyer[70] in Kent manufactured and sold many four-wheeled pedal-powered vehicles.

Pedals and cranked axles allowed the new bicycles to take advantage of the different radii of the wheels and the pedals (or cranks). This meant that the longitudinal distance travelled by the vehicle could be much greater than the circumferential distance travelled by the feet. Of course, there is a trade-off with the work produced reducing as the distance travelled increases – a minor factor when travelling on level ground but an important one when climbing hills. This is a simple application of the concept of gearing.

The use of gears to change the form of the mechanical output and the related concept of mechanical efficiency had both been widely utilised in 18th century wind mills, horse mills, water-wheels and clocks. Water-wheels powered by stream flows were commonplace and had required sound mechanical engineering skills at a relatively large scale. At a much smaller scale, mechanical clocks were common in Europe from the 14th century onwards and the intricate application of complex gearing systems had been well developed in 18th century clocks.[71] The clocks and watches were also pieces of great elegance and beauty. It has been argued that the clocks and the orderliness that they produced were major precursors of the Industrial Revolution.[72]

A subsequent approach to bicycle gearing was to have a large front wheel with pedals attached and a radius about equal to the rider's leg length. One rotation of the pedals on a much smaller radius arm moved the bicycle a distance equal to wheel circumference. Such vehicles were called *penny farthings* (referring to the relative size of the two wheels) or *high wheelers*. Of course, the torque was very low and the high centre of mass meant that the vehicle was very unstable, particularly when starting or braking.

Practical two-wheeled bicycles with pedals and cranks were produced and sold by the Michaux[73] family of Parisian carriage-makers in 1861. These vehicles – called *vélocipèdes* – were an order of magnitude better than their predecessors and revolutionised personal travel. However, they had iron rims and wooden spokes and so at high speeds they provided a very rough ride. Indeed, their English name quickly changed from *velocipede* to *boneshaker*. In addition, the pedals were still attached to the front wheels and so the gearing issue was unresolved.

The problem with penny farthings and other velocipedes was obviated by using a continuous chain going from a chain drive wheel at the pedals to a much smaller wheel on the axle of the rear wheel. Thus one turn of the pedal produced many turns of the driven wheel, which was further magnified by its much larger circumference. Of course, the price paid for this increase in speed is a drop in power as the force in the chain acts over the radius of the two wheels.

This major advance was applied by Henry Lawson[74] in London in 1877 when he produced the first bicycle with a chain drive and, thus, enhanced gearing. Bicycle-makers quickly adopted the new gear technology and the dangerous penny-farthing quickly became a thing of the past. Subsequently, gearing was widely used in the new cars as it allowed an internal-combustion engine operating at optimum engine revolutions to be used to produce a range of torque outputs, such as the low gear used for climbing hills. Most steam engines did not rely on gearing, as their power output could be altered by adjusting the steam supply.

Bicycles with many of the features of the modern bicycle were manufactured by John Starley[75] in Coventry in England from 1885 onwards. They used light tensioned-spoked wire wheels which had been invented by George Cayley[76] in 1803 in his search for a light flying machine and had been occasionally used in bicycles since 1867. Starley's uncle James was the first to use them for bicycles manufactured on a commercial scale. Previously wheels had used wooden spokes (Figs 2 & 3) which worked in compression and gave a much "harder" ride than the new wire spokes made of iron and working in tension.

This development of the bicycle wheel highlighted the need for better tyres than the solid wood, metal or leather ones then in widespread use. These relatively inflexible tyres produced uncomfortable travel, inhibited increases in travel speed, and led to unsafe riding. Rubber was known to be far more flexible, but it could not be formed into useful shapes. This drawback was removed when Charles Goodyear,[77] working in Springfield, Massachusetts, invented the vulcanisation of rubber in 1839. This led Robert Thompson[78] in Scotland to use solid rubber tyres on steam-powered road vehicles in 1867 and they were soon widely adopted. They were better than the alternatives but were still damaging to passengers, cargo and pavements.

Thompson had patented a pneumatic tyre in 1846, but could not find suitable material for its manufacture. The solution was to use an air-tight flexible tube inside the stiffer and tougher tyre. This approach was not invented until 1888 and was due to John Dunlop,[79] a Scottish veterinarian working in Northern Ireland. His motivation was to allow his son to ride his bicycle at increased speed over the cobbled streets of Belfast. A major later advantage of the pneumatic tyre when used with heavier vehicles was that the pressure on the road surface could not exceed the pressure in the tyre.

Dunlop's pneumatic tyre and tube resulted in a quantum jump in both user comfort and vehicle performance and their subsequent use on motor vehicles is discussed in Chapter 16a. Bicycle manufacture henceforth flourished, particularly in Coventry where the 1891 census recorded 4 039 people working in the bicycle and tricycle industry. It is interesting as this stage in the story to note that this period of British cycle production was occurring when, as described in Chapter 16, motor vehicle production was commencing in Germany and France.

By the end of the 19th century bicycles had become relatively sophisticated and bicycle design has subsequently only changed marginally. However, the continued market pressure for lightness and strength in many fields had technical spin-offs into the 20th century. For example, in 1870 Japanese manufacturers used bicycle technology to produce the rickshaw (or jinricksha) which is a light vehicle with two adjacent wheels carrying a transverse seat for one or two passengers and pulled by a person on foot or on a bicycle. By the end of the 19th century, Japan had about 200 000 in operation and was exporting 10 000 a year to other Asian countries. Their manufacture was a forerunner to Japan's wider industrialisation. They became a major form of urban transport in many of the world's cities.

Technically, it could be argued that the bicycle had a very slow birth, taking about a century for the simple concept to reach some degree of fruition. The first hurdle was to produce a device which would travel fast enough for users to realise the enormous gyroscopic stabilising effect of the angular momentum of the wheels – a point that had quite eluded physicists. The second hurdle was that the beneficial gearing effects of pedals were only slowly implemented – another simple concept that was only usefully understood in late retrospect. Wire spokes and pneumatic tyres were dependent on other developments and can be omitted from this catalogue of slowly crossed hurdles. Perhaps the bicycle was seen as such a simple device that it attracted little of the competitive incentives that occurred elsewhere during the 19th century?

As will be seen in latter Chapters, motor vehicle development also drew heavily on bicycle technology in such areas as ball bearings, tubular steel frames, differential gears and chain drives. Bicycle manufacture had also created a skilled work-force that would provide a source for future automotive mechanics. At the other end of the transport spectrum, when the Wright brothers were building the first aeroplane at the beginning of the 20th century, they also drew heavily on bicycle technology when producing chain drives, wheel components, and tubular steel frame design. These were all new technologies and new skills and their emergence in the 19th century is a central theme of this book. We will leave a discussion of that emergence – particularly how and when and why it occurred – until later Chapters of this book.

But the impact of the bicycle went beyond its associated technologies. Cycling became extremely popular for both work and recreation, across most levels of society. Consequently, cyclists organised themselves into active groups which in many countries became the leading advocates for better roads and relevant traffic regulations. For instance, in the USA the first bicycle club was formed in Boston in 1878 and a national body, the League of American Wheelmen, was formed in 1880, just three years after the first imported bicycle arrived in the US. The League became a highly effective lobby group for cyclists' rights and for better roads. Rights were important as bicycles were arbitrarily banned in a number of cities, such as Boston and Brooklyn, and notably in Central Park in New York, mainly as a consequence of conflicts with horse riders. Another battle was to convince the railway companies to permit bicycles to be taken on trains.[80] "*Better roads*" was a constant theme of cycle groups as even with good tyres, bicycles require relatively smooth surfaces and loose gravel was particularly hazardous. Thus by 1891, and well before the political impact of the car, New Jersey reacted to the bicycle lobby and provided legislative leadership by introducing effective state-wide taxation measures to fund road construction and maintenance.

In Britain the cyclists set up the Roads Improvement Association "*to press for better administered and better kept roads*".[81] Although cycling was popular, cyclists were not universally well regarded. They were said to be "cads" and "roughs" who scared horses and pedestrians and rarely paid municipal taxes.[82]

The transport impact of the bicycle was very different to that of the steam-powered road vehicles and railways which were coming into use at the same time (Chapter 9). Bicycles were cheap to buy and easy to maintain – far cheaper to own and operate than a horse – and they appealed to all sectors of society. As a mechanical device, they were about four times more cost-effective than walking.[83] And they could be used for both work and pleasure. Perhaps most importantly, they introduced people to the idea of independent personal mobility at speeds far greater than walking and without the imposed external constraints, timetables and fares of organised transport. Initially, their only competitors for road space were horse-drawn wagons and coaches, and much of that traffic had been captured by the new railways. Steam-powered road vehicles were still relatively impractical, weighing about 3 ton, requiring a driver and stoker, and consuming relatively large amounts of fuel and water (Chapter 9e). The bicycle not only addressed people's current transport needs but also created completely new transport markets.

It has been observed that in the 1890s "*for the first time since the rise of the railways the roads were crowded with through traffic, and, for perhaps the first time ever, the idea was born of travelling the road for pleasure*.[84] This latter remark is certainly true if it applies to the population in general – however there had been many previous examples of admittedly select individuals delighting in the occasional pleasures of road travel.[85]

Most of all, the bicycle introduced the general population to the idea of affordable, individual longer-distance travel for work and for pleasure. In a significant social and technical sense, people finally became masters of their own travel destinies.

PART 3

ROADS IN THE 18TH AND EARLY 19TH CENTURIES

CHAPTER 3A

BRITISH ROADS

The first two Parts of this book explored the development of wheeled vehicles over time and prior to the availability of useful mechanical power. The vehicles that did exist could only be used for transport if there were usable tracks, roads and water crossings.

European land transport in the period before the 19th century had relied largely on a road system inherited from medieval times and with only a few links to the earlier roads constructed and maintained by the Romans.[86] Roman roads are justifiably famous, but they had required major resources to build, operate and maintain. Once the Roman Empire dissolved, its roads were left uncared for and deteriorated rapidly under the influence of traffic, weather (particularly water), the invasive efforts of vegetation and abutting landowners, and a desire by local communities to be isolated from marauding outsiders.[87]

Thus roads in the Middle Ages were poor and most efforts to maintain them were misguided. Usable tracks mostly conformed to the demands of the local terrain and topography and existed with minimal human intervention. They predominantly served the local needs of foot traffic and served, or attempted to serve, the commercial, social and religious needs of local and regional communities.

The slowly increasing demand for travel after the Middle Ages placed growing pressures on the road systems. A further pressure was the increased fencing of private land as part of a process known as "*enclosure.*" A side-effect of this practice was that it prevented travellers from circumventing impassable pieces of road by using undamaged surfaces in adjoining fields.

Impassable roads were commonplace as only rarely was it possible to marshal the resources needed to maintain road surfaces, clear vegetation, construct cuttings and embankments, and build bridges. Organisations with some passing responsibility for the maintenance and operation of roads were usually weak and impoverished and so their efforts to provide adequate roads were rarely successful.

As the situation with roads worsened, England turned increasingly to roads managed and maintained by private groups and funded by imposing tolls on travellers. In much earlier times, travellers had been charged tolls to pass through gates in city walls or to cross bridges. For instance, there are records of tolls being collected in India in 320 BC. Tolls were imposed on the use of London Bridge when it opened in 1286 and remained in place for the next 545 years. However tolls on road lengths had less obvious value and required more sophisticated collection systems than bridges and city walls. Not surprisingly, such tolls were often hotly disputed by locals making short trips. The toll roads were thus called "*turnpikes*" due to the gates armed with long spears (or pikes) which were used to stop travellers and collect the appropriate toll.

Unsuccessful attempts were made in England in 1609 and 1622 to legalise toll roads with the first toll road legislation being an Act of 1661-3 which permitted tolls to be collected on the London end of the Great North Road. The main customers were malt wagons supplying London breweries. Evasion levels were high and the operating trust finally failed in 1733. Other toll roads encountered savage rioting by protestors. Despite this, the number of toll roads continued to increase. Each had its own piece of legislation and between 1751 and 1771 the British Parliament passed some 870 Turnpike Acts. This was the period of greatest growth in the system and it has sometimes been called the time of *Turnpike Mania*.[88]

As the turnpike system developed, British roads gradually improved and travelling became safer, easier and cheaper.[89] However, the growth in local and regional trade in the 18th century led to increased road usage and this caused significant damage many existing roads. Arthur Young was a respected commentator who toured England extensively between 1760 and 1800. He considered that half the main roads he encountered were good and a quarter were in moderate condition. His reports to the Board of Agriculture of road conditions for the remaining quarter of the road network were scathing.[90] Of the road from Preston to Wigan he wrote "*I know in the whole range of language no terms sufficiently expressive to describe this infernal road*."[91] A wide review of local records and reports for the period confirms the "*shocking condition*" of British roads during this period and observes that "*in winter months highways became impassable in a sea of mud*".[92]

By the 1820s Britain had about 40 000 km of turnpike roads, covering most major roads and many secondary roads. A formal census of toll roads in 1840 found that there were 1,116 toll road operators ("Trusts") managing 7,796 toll gates and controlling 22% of Britain's highways, with a concentration on the more heavily trafficked routes. A feature of the 19th

century turnpikes was that the northern industrial counties of Yorkshire and Lancashire accounted for over 30 per cent of the new trusts formed, whereas earlier trusts had had a stronger London focus, although there was also a major system from the port of Bristol to the mines and iron-works in the Severn valley.[93]

The development of the British road system in the turnpike era had occurred with little or no planning of the system at a national rather than a local level.[94] Nevertheless, the concentration on heavily used roads meant that the system came to reflect the economic imperatives of the time.[95] Placing a charge on road use had a range of social and local impacts, but from a transport aspect it brought the road system into the market economy at a time when markets were broadening in extent and impact. However, road development was massively interrupted by the era of steam-powered railways which began in the 1830s. Over a few decades turnpike companies drifted from profit-making to bankruptcy. Indeed, steam-powered railways had a major impact or roads and road transport throughout the world. This impact and the late 19th century response to it are explored in some detail in Chapter 11d.

The tolls charged were controlled by the relevant Act and hopefully had covered cost recovery and a return on investment. In 1842 Arsène Jules Dupuit[96], a major French economist, conducted a study of charges on toll bridges. His work led him to propose *marginal costing* for setting toll charges.[97] A marginal cost is the extra cost caused by a user and does not include any capital costs. It was rarely used on the 18th and 19th century toll facilities but remains today as a central tenet of the economic theory for costing any public service.

The 18th century roads set another very different precedent. The Romans had used their roads to control, administer and tax the various territories that they had conquered and made part of their Empire. Early in the 18th century Britain was having trouble with rebellious and warlike Scottish clans and in 1725 decided to adopt the Roman strategy and begin building roads into troublesome and previously inaccessible parts of Scotland. This was Britain's first step into national road-making.

CHAPTER 3B

ELSEWHERE IN EUROPE AND AMERICA

The French road system was dominated by the central role of Paris in the administration and control of the entire country and by the government's heavy reliance on a strong central army, conditions which did not apply in Britain. The system began in 1464 when Louis XI initiated a national courier system and this soon led to the first French national road between Paris and Orléans. Major roadwork occurred in the 17th century under Finance Minister Jean-Baptiste Colbert, whose projects included the Champs-Elysées which was built to service royal palaces on the Seine. In the 18th century there was extensive road-building to accommodate the needs of the aristocracy who also financed much of the work, although the labour was usually supplied by obligatory work impositions on the local peasantry. Indeed, such forced statutory labour is said to have been one of the prime causes of the French Revolution.

In 1716 France created a central road authority – the Corps des Ponts et Chaussées – and in 1747 a training school for road engineers, the École Nationale des Ponts et Chaussées. Jean Perronet was appointed head of the new school and had a major impact on the development of French roads. As a consequence of these various measures, the French road system at the end of the 18th century was the best in the world. It comprised some 50 Mm of constructed road, of which 40% was paved. The School became the model for engineering education in Prussia, Austria, Russia and the USA. For example the US Military Academy at West Point used the École as the model for formulating its own academic structure[98] and students at the US Academy were required to have a working knowledge of French.[99] Louis XV, who reigned from 1715 to 1764, is regarded in France as the "father" of its paved system of national roads.[100]

A continuing strong desire to protect and enhance the central role of Paris in the management of France and of its neighbours led to the continued development of a network of arterial roads. Quality was assured by the trained road engineers graduating from the École Nationale and by the need for the roads to be able to accommodate large marching armies.

The French Revolution began in 1789 and in 1791 the old aristocratic road funding source was abolished. Napoléon Bonaparte seized power in 1799 and in 1806 he reintroduced road taxes and in 1811 he strengthened the powers of the central road body. This new situation would allow France to avoid the road deterioration that occurred in Britain with the coming of the railways. The administration of roads in Napoleonic France had a major influence on the other European and American road systems, particularly that of Germany. With the technical support of Nicolas Céard,[101] Napoléon also built new roads to serve his imperial needs, most notably the trans-alpine passes using the Mont Cenis Road between France and Italy (1803-10, Fig. 17) and the Simplon Road (1801-5) between Switzerland and Italy. The Simplon has been described as the first work in Europe to surpass that of the Romans. It has been estimated that between 1804 and 1812 Napoléon spent twice as much on roads as on military fortifications, and more on bridges than he did on buildings.

Fig. 17 Mont Cenis Road in use by horse-drawn vehicles. Early engraving via Wikipedia Commons.

As in many countries, the coming of the train in the 1830s saw the management and funding of most roads returned to local governments which rarely had the cash or the incentive to even maintain existing roads (Chapter 11d). This was particularly the case in Britain, however the problem was avoided in France where local governments had gained the power to collect the taxes needed to maintain and improve their roads. Thus, in the railway era the French road system continued to be the best in the world. Towards the end of the 19th century this had a major influence on vehicle development. Although, as will be demonstrated in later Chapters, road vehicles powered by internal-combustion engines were invented in south-western Germany in the 1880s, by 1900 they were almost entirely being built, sold and used in France. Chapter 16b advances a number of reasons for this extraordinary shift, with one being that at that time only France had a road system suitable for use by the new cars.

German history in the 19th century is complex.[102] For most of 18th century much of Germany was part of the Holy Roman Empire although the north-east was part of Prussia, centred on Berlin. In 1796 the conquering Napoléon created the Confederation of the Rhine by amalgamating a number of German States and the Holy Roman Empire was subsequently obliterated in 1806. Compared with France, for most of the 19th century Germany was a weak federation of states with a much stronger structure of regional government and a more extensive system of road and bridge tolls. Indeed, the many administrative boundaries had led to a proliferation of toll-gates. This arrangement grew and survived until Prussia formed the North German confederation in 1871, following victory in the Franco-Prussian War. This grouping was to become the basis of 20th century Germany.

In 1816 in the era of Napoléonic control, a road construction and repair manual was issued[103] and in 1825 a Polytechnic School was established in Karlsruhe based on the French École Nationale des Ponts et Chaussées.[104] The processes of road management used in France were applied by many of the individual States within the federation.

Despite some road improvements, coach services were still slow and average travel speeds between centres such as Frankfurt and Stuttgart and Berlin and Hamburg were at about 5 km/h, or walking speed, and a day's journey would rarely exceed 120 km.[105] Napoléon during his period of control built only two roads in Germany. For obvious strategic reasons, they were east-west access roads from Wesel to Hamburg and from Metz to Mainz.

As in Britain, the introduction of steam railways in the 1840s put a massive dampener on any further major road development. The new railway networks took increasing amounts of the long-distance transport and drops in demand for road travel meant that many of the 38 German states removed road tolls from 1840 onwards. As the damaging effects of this policy became apparent the situation began to change and the German road network doubled in length in the last quarter of the 19th century.

The situation in most of Europe mirrored that in Germany. Rivers and canals, particularly in Russia, often provided the most reliable means of surface transport.

The American road system had developed to meet the local needs of European settlers, often using existing tracks and trails made by animals or the various indigenous peoples. In the 18th century there were also some roads built to meet militaristic needs. In 1802 Presidents Thomas Jefferson and George Washington were strong advocates of a National Road to aid western expansion and assist in holding the large new nation together.[106] This had been a matter of discussion since independence in 1786 but was brought to a head when Ohio was admitted to the Union. In 1806 the enabling Act was passed by Congress with strong support from President Jefferson and his influential Swiss-born Secretary of Treasury, Albert Gallatin[107,108]. Part of the funding was to come from a tax on land sales. The road was to run west from the navigable limit of the Potomac River at Cumberland in Maryland. Construction began in 1811 and when work stopped in 1841, it had reached Vandalia in Illinois – a road distance of about 1000 km. These dates of expansion and decline mirror the development of the new railways and closely align with similar national road building patterns in Britain.[109]

PART 4

THE 19TH CENTURY BRITISH ROAD SYSTEM RESPONDS

CHAPTER 4A

THE ROAD ENGINEERING PLAYERS

The four aspects that determine the effectiveness of a road between points A and B are:

1. The broad geometry of its horizontal and vertical alignments,
2. The presence of local geometric impediments such as sharp curves and steep slopes,
3. The surface of the road,
4. The width of the road.

The specific requirements under each of these headings became much more challenging with the introduction early in the 19th century of the new coaches described in Chapter 2. Indeed, it became obvious that the new requirements could only be met by external engineering interventions that would not be possible under the then existing systems of piecemeal local road management. This chapter explores the response to that realisation.

An important harbinger of the required changes was John Metcalf who built some 300 km of roads in the north of England in the second half of the 18th century. He emphasised common-sense principles related to drainage and materials that had eluded road builders for the preceding millennium. He also invented a method for "*floating*" roads across deep swamps – a technique late copied by the Stephensons in constructing the Liverpool-Manchester railway (Chapter 10a). Elsewhere, I have published a review of Metcalf's many extraordinary talents and achievements.[110]

The two major 19th century change-makers were Thomas Telford and John McAdam. There are some common threads in their backgrounds and both owed an inheritance to Metcalf. Telford (Fig. 18) was born in 1757 in Langholm in the south of Scotland "*in one of the most solitary nooks of the narrow valley of the Esk*".[111,112] He was born to a poor family as his father was a shepherd who died within six months of Thomas' birth. Telford was taught at the local parish school. He had aspirations to be an architect but his circumstances required him to begin his career as an apprentice stonemason, repairing local houses and farm-buildings. Before long he was building churches and manses. In 1780, he left Langholm for Edinburgh and was soon producing beautiful architectural drawings. Two years later he moved to London where his drawing skills quickly gained him work with a leading architect. Within a few years he was building harbours, canals and bridges and soon became one of the outstanding engineers of his era.

Fig. 18 Thomas Telford (1757 – 1835). Contemporary engraving by Samuel Lane.[113]

During his career he successfully tackled an astonishing range of infrastructure-related endeavours. He became a master bridge designer cum builder and an innovator in traditional masonry, in the newer cast irons, in suspension bridges, in tunnels, in harbours, in canals and in roadmaking. He designed from intuition rather than by calculation, yet many of his technically challenging structures are still standing. His bridges were innovative, handsome and stylish (e. g Figs 24, 25, 27 & 28). He built mammoth canals, harbours and wharves where none had been before. He created new methods of construction and introduced a scientific approach to road layout. He built numerous new roads in difficult terrains.

He built tunnels for roads and canals and some extraordinary canal bridges. He worked throughout Britain and in Poland, Sweden, Canada and India. He was the first County-Surveyor of Shropshire. He was active in land reclamation and the development of fisheries. He was a great project manager and organiser and readily gained and kept the confidence of those who knew him – from workmen to politicians. He was admired by his peers and the poet Robert Southey was a great friend who gave him the punning and appropriate nick-name, *the Colossus of Roads*. He was a founder of the Institution of Civil Engineers, a body that still flourishes today. He was clearly a man for and beyond his times and was certainly one of the major change-makers during the first half of the 19th century.[114,115,116,117]

Telford's involvement in roads began in 1801 when he was commissioned by the British government to report on the transport measures needed to halt the depopulation of the Scottish highlands. He subsequently applied his exceptional engineering skills to the selection of road alignments and to road-making. The resulting works were substantial and effective, but relatively expensive. A fine example is the Glendaruel Road which connected Rothesay and Inveraray in the Scottish Highlands. The 30 km road and 10 m span bridge opened in 1812 and reduced the previous trip length by 160 km.[118] A later example from 1812 is the 150 km road from Glasgow to Carlisle, of which about 100 km was on a new alignment.[119] From 1820 to1830 "*an immense amount of road survey work was done under Telford's direction*".[120] The largest of these works would have been the Great North Road which was to provide a mail coach route over the 600 km from London to Edinburgh. Telford began surveys for the route in 1827. However, the only portion built to his design was the final 20 km south of Edinburgh as Telford's attention was then diverted by the government to the design and construction of the Great Irish Road. This project is discussed in Chapter 4c.

McAdam[121] was also Scottish, coming from Ayr which is about 70 km north-west of Langholm (Fig. 19 &[122]). He was born a year earlier than Telford and died a year before him. McAdam's father had been a substantial property owner in Ayrshire and his son initially worked in the local commercial sector and then for 13 years with an uncle in New York. After the American War of Independence he returned to Scotland in 1783 and bought a country estate in Ayrshire. He had his first brush with roads when he was appointed a trustee of the Ayrshire Turnpike in 1787. He was almost bankrupted by a business venture and in 1798 he was working at Falmouth in the south-west of England servicing the needs of the British Navy. He subsequently estimated that for the next 16 years his private affairs required about 3 000 km of road travel annually. He became a very astute and critical observer of the condition and management of the roads that he encountered during these travels.[123]

Fig. 19 John Loudon McAdam. Contemporary engraving by Charles Turner. Wikipedia Commons.

McAdam's immediate contribution was to road management and his lasting contribution was to pavement technology. Turning first to his contributions to road management, in 1801 he moved to Bristol where his side interest in roads soon led to his appointment as a trustee of some local turnpike trusts. In this major mid-life career change he proved a good manager and organiser and a tireless publicist and advocate. In 1816 he was appointed General Surveyor of the relatively large (250 km of road) Bristol Turnpike Trust and within three years he held similar positions with 33 other Turnpike Trusts. He did not build many new roads and certainly maintained and repaired far more than he built. The title

Surveyor implied a general management role rather than possession of the specific technical skills required of a modern land surveyor.

Scottish politician John Sinclair's Select Parliamentary Committee appointed by the House of Commons to inquire into British roads is discussed in the next Chapter. McAdam appeared before the Committee in 1816 and whilst the evidence he gave was precise and articulate it was also extremely blunt and critical of the current situation. His biographer described it as "*forthright and acidulated*".[124] McAdam did not restrict its impact to Parliament but also used it as the basis of a published and widely-read guide to all aspects of road-making that went through nine editions between 1816 and 1827.[125] Amongst his acerbic public remarks are the following (paraphrased):

Trustees of private roads are narrow and jealous. Their accounts are obscure, confused, perplexed and dilapidated. The defect in their road-making lies in the want of science. They cause a great waste of public money.

Road engineers and surveyors deceive the public and delude the trustees. They are elected because they can measure; they might as well be elected because they can sing. Their want of science has gone hand-in-hand with improvident expenditure. They are without energy or character.

Public servants: worthless.

Between 1816 and 1826 McAdam and many of his family were managing about 80 toll companies operating 3000 km of toll road. In 1825 his son James[126] was controlling 40 companies and 1400 km of toll road. In following year James was appointed Surveyor-General for the Commissioners of the Metropolis Turnpike Roads North of the Thames. His major project was to consolidate London's fourteen separate toll companies and their 200 km of busy road into a single entity.[127] The work of Commissioners peaked in 1835 and disappeared in 1872 as a consequence of the dominance of railway era issues discussed in Chapter 9. An equivalent role did not reoccur in Britain until 1937 when the Ministry of Transport was given responsibility for Main Trunk Roads.

James McAdam's principle biographer describes him as belonging to:

"*the generation of reformers who took the administration of 18th century England by the scruff of its neck and shook it into the beginnings of its Victorian shape. Not for him the part-time amateurism, combined with patronage of the old order. He believed in full-time professional* *competence and financial control, founded on accurate facts tirelessly collected, tabulated and assembled.*"[128]

Telford and the McAdams soon became major forces in road construction for the burgeoning toll road industry described in Chapter 3a. Whereas the McAdams had moved increasingly towards management of the road system, Thomas Telford's work always had a strong technical orientation, although he was the instigator of the engineer-contractor relationship and the role of the engineer as the independent arbitrator between owner and contractor – a system which still underlies today's private sector contracting. And whereas McAdam's emphasis was on the pavement, Telford examined – theoretically and experimentally – the links between road alignment and vehicle performance and tested the hauling characteristics of animal-drawn vehicles.

Turning now to John McAdam and Thomas Telford's contributions to pavement technology, between 1816 and 1830 McAdam developed an alternative road-making method that produced pavement outcomes that were at least as good as Telford's, and were far cheaper. At the time there was fierce competition between the two and their supporters, but McAdam's method has stood the test of time whereas today Telford's is rarely, if ever, used.[129] Pavements made by McAdam's method came to be called *macadam pavements*, and this name has also lasted to the present day.

McAdam's method was a major paradigm shift in pavement making.[130] For many it was also counter-intuitive and led to major controversies, some of which persisted for another hundred years. McAdam found – possibly by observing the effect of iron-rimmed truck wheels on broken stones in a mason's yard – that good pavements could be made by placing and compacting layers of broken, sharp-edged stones on firm, well-drained lengths of natural formation. Initially the pieces of broken stone – less than 30 mm in size – were produced and compacted by hand. Limiting the stone size was a key part of his method – previous practice had been to use larger, randomly-sized pieces of unbroken stone to build and repair road pavements. McAdam chose his stone size to be smaller than the width of the rims of the wheels of carriages and coaches and of the hooves of horses and oxen. Passing traffic therefore compacted the stones into a layer rather than displaced them from the layer. The poor performance of rounded river gravel – particularly in the London area – had also emphasised the need to use broken stone.[131]

McAdam's practical demand was that the stone had to be small enough to fit into the stone-breaker's mouth. Telford required it to be smaller than a hen's egg when used in his surface layers. The quick adoption of the McAdam approach was due to three main factors:[132]

1. it used local unskilled labour and, often, local stones,
2. it produced good road surfaces, and
3. it cost about a tenth of the cost of the Telfordian alternatives made from stone blocks fitted together to become a piece of horizontal masonry.

The McAdam approach created a great deal of manual work, which was very convenient in a time of unemployment exacerbated by the return of British soldiers from their triumphs at the Battle of Waterloo. Steam-powered devices described in Chapter 13b for quarrying, crushing, carrying and compacting stone would subsequently replace these manual processes. The use of compacted layers of relatively small, sharp-edged, cubical stones became a basic component of road pavements, and remains so today two centuries later.[133]

The paradigm shift was the realisation that a horizontal layer of broken stones when compacted and restrained from lateral expansion, would act as a coherent layer under the vertical and horizontal forces caused by passing traffic. This was primarily due to mechanical interlocking of the sharply-edged, angular, stones. The method was quickly adopted by the new railways and used as the layer – called *ballast* in railway practice – between the sleepers and the natural formation (Chapter 9c).

Another strength of McAdam's approach was that he perceptively avoided the various fictions of earlier road-makers and emphasised many of the lessons of the preceding century of improved road making and maintaining and of roadside drainage. For example, an excellent summary of good practice at the beginning of the 19th century is given in Bramley's 1805 Road-makers Guide[134] and McAdam's subsequent advice is consistent with Bramley. Bramley and McAdam remain relevant today. On page 3 Bramley succinctly reminds his readers that "*the first object desirable in roads is undoubtedly the facility of conveyance; the second is to continue the duration of that facility*."

The 1820s was certainly a time of major activity for Britain's two major groups of road creators, McAdam and his family, and Telford and his many colleagues, such as Parnell and Macneill. McAdam's criticism of the status quo in 1816 has already been noted. Twenty years later Telford's friend Parnell was to write:[135]

> "*The business of road-making in this country has been confined almost entirely to the management of individuals wholly ignorant of the scientific principles on which it depends. It has received, till very lately, little attention from the scientific world; so little that the primary and*

indispensable objectives ..... have not, till within a short time, been recognised and explained by any scientific rules whatever."

Given the enmity between the Telford and McAdam camps, McAdam was probably included in "*wholly ignorant*" category.

Even after the innovative work of Telford and the McAdams, road pavements bore only a passing resemblance to their current-day counterparts as tar, bitumen, asphalt and cheap cement were still half a century away. Roman paving practice was well known, but very expensive to implement.[136] One age-old paving solution was to surface the streets with stone slabs (flag stones) – a process that was expensive and still required a firm base on which to place the slabs. Over the centuries there had been increasing use in large towns of smaller cubes of stone called cobbles or setts. Although they provide a rougher and noisier ride, they were favoured by transport operators as the animal hooves could obtain a better grip on the slime-covered pavement. Wooden blocks were also used, however they wore more quickly than stone and were not very durable – particularly European blocks made from softwoods such as Baltic pine.[137]

Thus road pavements at the beginning of the 19th century were scarcely suited for existing traffic, let alone for what lay ahead. Cobble stones and wood blocks were expensive and only partially effective. On heavily trafficked roads and streets macadam surfaces represented a major improvement but as wheeled road vehicles became increasingly more commonplace during the course of the 19th century there was a growing demand for pavement surfaces that were smooth enough for comfortable and quiet travel, rough enough to provide friction for the operation of tyres and hooves, readily washed and drained, and strong enough to resist the vertical pressures imposed by tyres and hooves. In towns, a more coherent surface was needed to resist the more concentrated effects of animal hooves and narrow wheels and to make it practical to drain the road surface and remove the daily layers of excrement produced by the animals used for transport and the waste from adjacent houses. The problem was a significant one for most towns and worsened as traffic numbers increased.[138]

Chapter 17c discusses the curious coincidence that saw two major transport problems – pavements and vehicle power – solved by one new product. Consequently, by the end of the 19th century the old pavements were being successfully replaced by asphalt and concrete.[139]

CHAPTER 4B

THE BROADER POLITICAL AND ADMINISTRATIVE PROCESSES

The processes behind the 19th century work of Telford and McAdam discussed in the previous chapter deserve further examination. Arthur Young's comments on the state of British roads in the second half of the 18th century have already been given in Chapter 3a. Some of his work was commissioned by the Board of Agriculture and initially Young was the Board's Secretary. The Board had been created in 1793 by the British Parliament following pressure from agricultural interests and has been described as "*a curious body with no bureaucratic function or authority – composed of enthusiasts – and supported partly by public funds and partly by private subscription*".[140] It gradually gained greater official status. The first President of the Board was Sir John Sinclair, a Scottish politician from 1780 until 1811 who had a passion for collecting statistics relating to issues he was pursuing – which were many and varied.

Roads were a major concern of the Board as they were vital to the farm-to-market activities of the agricultural community. Scottish land-owners and agriculturalists had a particularly strong interest in an improved system of regional roads as taking their products overland to markets in the south of England required passing through a myriad of toll gates on poorly connected roads. The Board lobbied for a Parliamentary Committee to be appointed to inquire into the administration of British roads and this occurred in 1808.

The British were possibly influenced by a far-sighted 1808 report by the US Treasury Secretary Albert Gallatin. It was perhaps the first formal economic assessment of roads and led to a comprehensive plan for a national system of roads and canals in the USA. However, constitutional arguments over State and Federal responsibilities and then the advent of steam power meant that very little of his plan was implemented.[141]

In 1811 John McAdam's work had come to Sinclair's attention and Sinclair in turn had put McAdam's efforts before the Parliamentary Committee. McAdam was soon acting as an Advisor to the Committee.

The Committee met over the next 22 years, said much but had little impact.[142]

Strong official pressures for road improvements also came from the Post Master General (PMG) and his obligation to oversee the performance of the Post Office and the Royal Mail Service. From 1780 these organisations had had a monopoly on letter-carrying but other objects could be distributed by contractors. The resulting mail delivery service was highly organised and meticulously managed, using the best coaches and ancillary services as it criss-crossed the country. Consequently, it was very well regarded and increasingly relied upon throughout Britain. Its careful map-making and travel-time record-keeping made it clear to all which roads were impediments to reasonable communications. The PMG reported directly to Parliament and made sure that problems with mail delivery message were heard at the highest levels.

The postal service relied totally on the road system and had simultaneously shown the benefits of a good road system and the costs of a bad one. In the 1750s the coach trip from London to Bristol took two days. The new postal service reduced this to 38 h. As discussed in Chapter 1e, a major initiative for service improvement came from John Palmer of Bath who in 1782 organised a London-Bath-Bristol horse-drawn coach service which made the journey in an extraordinary 17 h. Palmer and others then became contractors to the PMG. By the 1830s the journey was taking less than 12 hours which suggests an average speed of under 20 km/h. Today's road distance is about 200 km. The need to travel in stages, stopping at inns to change horses and refresh passengers, would suggest peak speeds of over 20 km/h.

The British mail coach system was copied in many countries. In northern Europe, Prussia used it as the model for its Schnellpost implemented between 1818 and 1822, with services extended beyond Prussia to Paris in the west and St Petersburg in the east.[143]

Britain came to increasingly depend on its efficient postal service. For example, the political pressures to construct the Great Irish Road (discussed in Chapter 4c) to aid the administration and control of Ireland began with PMG reports in 1808 listing the delays caused by the existing road component of the London to Dublin postal service. Similarly, it was the PMG who appointed Thomas Telford in 1827 to begin surveys for a new Great North Road between London and Edinburgh. It has been estimated that in Britain "*movement between major towns was four or five times as fast in 1830 as it had been in 1750, and almost everywhere else it was at least twice as fast*".[144] The trip between London and Edinburgh had taken 10 days in the 1750s, weather permitting. In 1836 it took less than 2 days.[145]

The benefits introduced by the PMG flowed on to private travel which was provided by many private sector operators. In Britain, the largest private coach company in the 1830s used 64 coaches and 1800 horses. In 1836, there were 342 scheduled daily coach departures from London. Of these, 275 were provided by just three companies.[146]

Thus, British roads had steadily improved from the 1780s through to 1840, when the new railways to be discussed below provided a dramatically better way of distributing the mail. After that date investment in roads went into rapid decline and service levels dropped back to those of earlier times (Chapter 17).

CHAPTER 4C

THE GREAT IRISH ROAD

Chapter 1e noted that by 1820 the Royal Mail coaches could provide a faster service than riders on horse-back. At its peak the Mail's horse-drawn services were operating at about 25 km/h – fast, but soon to be well exceeded by the steam-powered Rocket discussed in Chapter 10a. Nevertheless, from as early as 1810 it was obvious to the Royal Mail management that the increasing demands for its services were now being limited more by the poor roads than by the coaches and their teams of horses.

Britain had annexed Ireland via the Act of Union in 1800 which made it legally part of Great Britain in 1801. In Ireland, this led to a serious rebellion in 1803. Irish landlords who required both political stability and access to the London markets argued strongly that good communications between London and Dublin were vital. They were supported by military leaders, civil servants and Irish Members of Parliament, all of whom needed quick access between Westminster and Ireland. The shortest relevant crossing of the Irish Sea was the 100 km journey between Dublin and Holyhead in north-west Wales. The land journey from London to Holyhead is about 460 km. An official weekly postal service between London and Dublin had been operating on the route since 1576.

The road route from London was to the north of Birmingham and then to Chester. This route followed the line of Watling Street, which dated from at least Roman times. After Chester the route turned west into Wales, where it encountered some difficult terrain. In 1810 Parliament, via the Post Master General whose service relied heavily on the road, asked Thomas Telford to report on the road as it was taking at least 41 hours for the Irish mail coach to travel from London to Holyhead. Later that year Telford delivered a "*thorough and melancholy report*" on the condition of the route.

In 1815 Parliament established a Holyhead Road Commission to be funded partly by government grants and partly by tolls on users. It was the first and only time that such an approach was used in Britain. Parliament appointed Telford to plan and build the road. In his autobiography,[147]

Telford notes that what followed was "*fifteen years of incessant exertion*". Meanwhile, in 1819 a steam-powered boat service was introduced on the sea-crossing between Holyhead and Dublin and improved the reliability of that aspect of the service (Chapter 9d).

Telford's selected route went from London to Coventry, south of Birmingham and then on to Shrewsbury, to Bangor on the Welsh coast, and thence to Holyhead. There were 10 independent toll operators between London and Coventry, 7 from Coventry to Shrewsbury and one from there to Holyhead. The route out of London followed the current A1 to Finchley, the A1000 & A1081 to St Albans, the A5183 to Flamstead, a long length of the A5 to Weedon Bec, the A45 to Coventry and Birmingham, the A41 to beyond Wolverhampton, the A464 to Telford, mostly the A5 to Bangor and the A55 to Holyhead.[148] Indeed, about 40% of the A55 is still on Telford's alignment.

Surveys began in 1815 and construction commenced in 1819, concentrating particularly on five difficult lengths of the route in Wales. A major task for Telford was the construction of a suspension bridge over the Menai Straits, near Bangor in Wales and about 30 km east of Holyhead. The bridge is discussed in Chapter 5c. The route passed through some countryside which was very difficult for road-making as it included major bogs, ravines, rivers and precipices (Fig. 20). Telford did not merely tackle this task as a construction challenge but he also introduced many new skills into selecting an alignment which optimised the combined costs of construction and subsequent operations. This was the first time such an approach had been adopted and included experiments to measure the performance of animal-drawn vehicles on various grades and types of road surface. The work of Telford and his colleague John Macneill included measuring the tractive haulage forces that horses could produce in various circumstances.[149] A number of the principles that he established for selecting grades and alignments remain in use today.

The maximum grade of the new road was about 4.5%, which allowed horse-drawn coaches to travel at an "easy" trot. This was the dominant design factor in the rugged terrain and its practical achievement required considerable engineering skill. As coach speeds would not exceed 25 km/h, horizontal alignment was much less critical, although the right-angle bends of the older legacy alignments were avoided. In organised races in the 1820s, some coaches reached speeds of about 28 km/h during a 30 minute trial.[150]

The route had been gradually improved from the 1780s and was known as the Great Irish Road in 1805.[151] With the creation of the Commission it was sometimes called the Parliamentary Route. When the entire route was completed in 1830, mail times dropped from 41 h (about 10 km/h) to 28 h

Fig. 20 Telford's road (see wall on the left) as it emerges on the Bangor side of the Nant Ffrancon pass. Notice the gentle gradient. This photo is taken from the old pre-Telford lane, which, unlike the A5, has a steep gradient to the top of the pass. Words and the photo by David Ackerley, www.cyclingnorthwales.co.uk.

(16 km/h). It was nevertheless the last British investment of public funds into major road-making for almost a century. What happened to cause this hiatus? In short, following the steam railway developments described in later Chapters, neither the Royal Mail nor Parliament could see any virtue in investing further public funds in providing a service which was markedly inferior to that now being supplied at no government cost by the private sector.

Telford and his colleagues remained of the view that steam-powered vehicles could be produced which could operate more effectively on the Great Irish Road and other similar roads than on the railway equivalents. In 1833 they financially launched a so-called "Steam Company" to put their views into competitive effect. The Company had no success and disappeared in the following year – not coincidentally the year of Telford's death.[152] By this time, much of the traffic on the eastern portion of the road had been lost to the new railways. In 1850 the railway service reached Holyhead and provided dramatically reduced travel times.

Consequently, in 1851 grass was reported growing on the road pavement near Anglesey.[153]

One of the purposes of this book has been to explore how far ahead the great change-makers saw – whether they were solving current problems or seizing future opportunities. Telford and McAdam certainly qualify as great change-makers, however there is no evidence that they saw the distant future with any clarity. McAdam made much better and cheaper roads to cater for the steady increase in horse-drawn traffic. Telford saw the need to transport heavier loads more frequently and addressed this by better road alignments and a more extensive canal system. Steam-powered railways were on someone else's horizon.

PART 5

MAKING AND USING IRON

CHAPTER 5A

MAKING IRON

The role of iron in the lead up to the 19th century has already been discussed in Chapters 2 and 3. It will become apparent in the subsequent Chapters that the availability of large quantities of good quality iron and steel, and of the machinery to work them into useful forms, were major factor in the Industrial Revolution. Indeed, most of the innovations of the Revolution had iron as an essential and common component. It is therefore appropriate to spend some time examining the key aspects of the commercial development and application iron.

Before the Industrial Revolution, wood was the dominant industrial material. It was used for constructing buildings, bridges and vehicles. It was burnt to provide heat for buildings and boilers. Although there were copious supplies of wood in the New World, the associated deforestation in Europe during the 18th and 19th centuries was bluntly demonstrating that wood supplies were exhaustible. There were alternatives to wood. For instance, in some applications where compressive forces dominated, masonry could be used and the time before the Revolution is well remembered for its magnificent masonry cathedrals, bridges, castles and manors. However, in most cases the real alternative to wood was not masonry but iron which was much stronger and stiffer than wood, although initially also much rarer and dearer.

Although iron was stronger than existing materials such as wood and masonry in compression, its major asset was its far greater strength in tension and shear. Only some vine species remotely rivalled its tensile strength. Nevertheless, poorly made and/or treated iron can be deceptively brittle and this led to many disastrous structural failures, as will be recounted in Chapter 5d.

Iron is the commonest of the metallic elements, and the sixth most common of all the elements. It bonds strongly and readily with oxygen and so the commonest natural forms of iron are its two oxides – *hematite*, Fe_2O_3 and *magnetite*, Fe_3O_4. Both oxides are widespread and when found on a large scale and high concentration in the ground, are known as *iron ore*. Hematite has a higher iron content than magnetite and most iron ore is predominantly hematite. Of course, both ore types will contain many other elements and compounds, and these are generically labelled as *impurities*.

To produce iron from iron ore requires the removal of the oxygen, a chemical process known as *reduction.* This needs temperatures which are sufficiently high for the oxygen to preferentially combine with carbon rather than iron, forming gaseous carbon monoxide or carbon dioxide.

Smelting is the process of separating a metal from the ore in which it normally exists in nature. The first man-made source of iron developed in association with copper smelting, as minor quantities of iron are a residue of the copper smelting process.[154] Copper has a relatively low melting point (1085 C) and smelting it became practical in about 4000 BC. The temperatures achieved in these early copper furnaces were well below the melting point of iron (1537 C) but were sufficient to permit reduction to occur. The resulting iron was mixed with other residues (called *slag*) and this iron-slag mixture formed as spongy small pieces called *bloom* or *bloomery iron.* The bloom was later softened by reheating and then hammered repeatedly to physically separate the iron from the slag. This working process is called *wroughting* and the resulting material is *wrought iron*. The product was sometimes called *malleable iron*, as wrought irons are relatively soft. Wroughting also made it possible to produce larger iron pieces by hammering smaller pieces together. By about 1000 BC blooms with low slag contents were being produced by using ores (such as iron sands and meteoric iron) with lower impurity levels.

In early iron-making the carbon required for reduction of the iron ore was usually supplied by *charcoal* which was traditionally made by heating wood in the absence of oxygen. The iron ore was heated by a charcoal furnace fire with the charcoal providing both heat to release the oxygen from the iron oxides to produce iron and also the carbon needed to then carry the oxygen away as one of the carbon oxides. Inevitably, some of the carbon atoms remain the mixture. Most join with iron atoms to form a molecular chemical compound known as iron carbide (Fe_3C) or *cementite*. It is present in irons with more than 2% carbon. It is strong but brittle and so may have a deleterious effect on an iron's performance. Some of the carbon is also caught within the *body-centred cubic* crystalline structure of the cooling iron. These carbon atoms are physically squeezed into the very limited space in the iron crystal, rather than as iron carbide molecules.[155] This structure has a carbon content of only 0.02% and is called *ferrite*. It is more ductile but weaker than cementite.

Iron-makers soon found that the best furnaces used the solid materials – ore, fuel, fluxes – raised sufficiently high in the furnace to allow air containing oxygen to enter from the bottom and pass upwards as it was heated. The process became common in the first millennium BC when the use of bellows to supply the upward blast of air – and thus more oxygen – to the fire, allowed temperatures of up to 850 C to be achieved and maintained. These blast furnaces continued to dominate iron production with the only major improvement occurring in the 14th century when water-wheels were used to power the bellows.

Iron ore, particularly iron sands, could be reduced at temperatures of 850 C but the iron still could not be separated from all the other materials present in the ore and in the furnace. The iron also contained many bubbles of gas which had been unable to escape from the cooling iron. The lumps of iron produced by this means are known as *pigs*. It was found that the best irons for wroughting were produced when the pigs were then heated from 850 C to about 1200 C.

By prolonged and skilled working, it was possible to flatten the bubbles in the iron in a pig and separate it from the slag. However, such iron was still relatively soft and prone to rust (that is, to re-oxidise) and could not maintain a sharp edge. The softness was diminished when some of the carbon from the charcoal used for heating was then diffused into the surfaces of the piece of iron. The percentage of carbon by mass required to harden the iron is very small at about 0.3%. At these low levels the carbon atoms are soluble within the molten iron. We will later discuss why such small addition can have such a large effect. Hardness at the surface has many advantages for applications where wear resistance or sharp edges are critical, but carbon-induced hardness in the body of a piece of iron leads to brittleness under tensile stresses.

In 1665 Robert Hooke had written in his Micrographia, in the manner of an alchemist's recipe for gold, "*Iron is converted into steel by means of the incorporation of certain salts, with which it is kept a certain time in the fire.*"[156] In practice, the words *steel* and *steeling* were used for irons where a controlled carbon regime was introduced at the surface of the material. Today, the process is often called *carburising*. Steeling was employed for making weapons and tools for cutting, picking and hammering.

Another surface hardening process developed at about the same time was *quenching*, in which the hot iron is plunged into cold water. The rapid cooling creates a much harder microstructure. Unfortunately, this hardened structure is also relatively brittle and a compromise between hardness and crack-resistance is often required. This can be produced by further wroughting and by a process known as *tempering* in which the iron is reheating to up to 500 C and then cooled.[157]

Strictly, the term *iron* should be used to describe the crystalline material formed by iron atoms and *steel* would be some uniformly distributed combination of iron and carbon. However, in current practice *iron* is used to describe materials produced from iron ore and *steel* for irons in which impurities from the ore have been minimised and the carbon content reduced to a controlled low level to balance strength and ductility. This realistic definition dates from the 19th century. Making large quantities of steel requires the ability for carbon contents to be carefully selected and controlled. The resulting steels are a mixture of ferrite and lamella cementite and may have a carbon content as low as 0.1%. The inter-crystalline lamella and the distorting effect of carbon on the internal crystal structure of ferrite can still exert significant influences on a steel's properties.

As furnaces further improved and higher temperatures were achieved, the heated material was able to be poured into crucibles to produce small pig iron *ingots*. Larger pieces were produced by heating the small ingots and then hammering them together. Many of the subsequent hammering and forging processes not only shaped and melded the ingots, but also aligned defects in the ingots into a common plane. This produced products with greater strength parallel to the plane, but with a lesser strength normal to it.

In about 700 BC the Chinese had further improved furnaces and were able to produce temperatures high enough to melt iron. There was a major side benefit of the new furnaces. As the iron absorbed carbon, the melting point of the material dropped from over 1500 C to 1150 C. This made it feasible to pour the material into moulds. This *cast iron* had many practical advantages over wrought iron, however it typically has a much higher carbon content at about 4% which leads to it being more brittle than wrought iron. Cast iron was not produced in the West until about 1500 AD.

Another stage was required to make the new product malleable and at least as useful as wrought iron. The furnace atmosphere contained little free oxygen. By subsequently placing castings in air in a charcoal-fired hearth at about 900 C, the oxygen in the air removed much of the carbon from the skin of the casting, producing an iron product with a useful shape and a malleable skin. The process is sometimes called *roasting*. It enabled useful tools such as knives, swords and plough-shares to be produced in *finery forges* in much larger quantities than had been possible with wroughting. When steam-power became available it allowed the use of much larger and more effective forges.

The process to develop useful quantities of useful iron and steel was thus slow and ponderous and its improvement stretched over many millennia. Nevertheless, the potential of the product was clear to many and so the effort continued, slowly but persistently.

CHAPTER 5B

EARLY APPLICATIONS OF IRON

These first iron-making processes could only produce relatively small pieces of iron which were used[158] for nails, tools, weapons and vehicle parts (Chapter 2). Cast iron was wrought into chain links and these were used for load-carrying chains in suspension bridges, beginning in China in about 400 AD. This was a major achievement as it allowed inherently brittle cast iron to be used to carry tensile loads. For greater tensile strength and reliability, the Indians later used links made from coils of drawn iron wire. The first structural use of iron chains in Europe was in 1237 AD on the Stiebende Bridge in the St Gotthard Pass.

In about 500 AD Indian craftsmen were able to produce the famous Iron Pillar at Qutub Minar near Delhi. It is composed of pieces of iron wrought into an imposing column seven metres tall and weighing six tons. The iron is relatively pure and the Pillar still stands today, black and unrusted.

Another early structural use of small pieces of cast iron was in making ties to lock together masonry blocks (Fig. 21) in arch bridges with high span-to-rise ratios, as in the wonderful Zhaozhou bridge at Zhaozian in Hebei Province in China (Fig. 22). The bridge was built in 590 AD and remains in service today.[159]

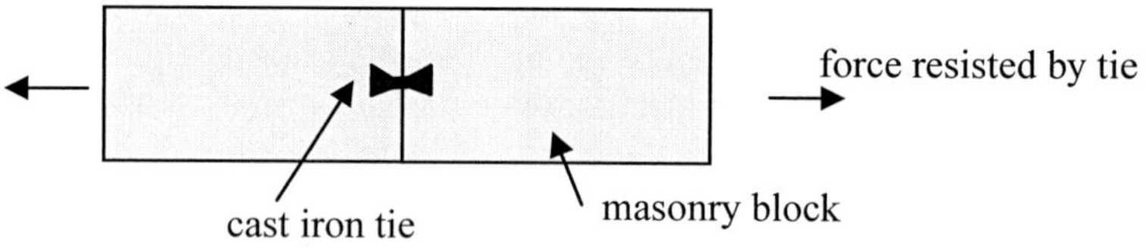

Fig. 21 Cast iron tie joining two masonry blocks, as used in the Zhaozhou bridge. See also Fig. 22.

The success of these applications encouraged improvements in iron-making. As discussed in Chapter 5a, the major need was to build furnaces able to produce very high temperatures. Medieval Europe had an abundance of trees to provide charcoal for the furnaces and water to turn water-wheels and provide power for the blast bellows. These factors allowed larger furnaces to be built and higher temperatures reached. This permitted greater quantities of iron to be produced and cast in sand moulds. Such large scale iron-making began at Ferriere in Italy towards the end of the 15th century. In the 16th century the demands of war pushed production further and cast iron cannons were a major product of the many new furnaces.

Fig. 22 Zhaozhou bridge at Zhaozian in Hebei Province. The iron ties explained in Fig. 21 are clearly visible. Photo by the author.

By the 17th century the fuel demands of the growing number of furnaces meant that wood was now in increasingly short supply. The supply problem was worse in countries with cold climates where wood was also widely used for domestic heating and cooking. The fuel shortage limited the production of iron. Although coal was readily available and widely used for heating, its impurities made it unsuitable for iron-making in the furnaces then in use.

The break-through that permitted coal to be used for iron-making was due to Abraham Darby[160] at Coalbrookdale in Shropshire in England early in the 18th century. It opened the door to the Industrial Revolution and many have said that Coalbrookdale and its surroundings were the birthplace (or the cradle) of the Revolution. How the once-fertile

Shropshire land came to be known as the Black Country will be explored in Chapter 6.

Rather than burn wood or charcoal or coal, Darby used *coke* to fire his furnaces. The coke was obtained by the burning of local coal in a low oxygen environment in a technique based on a process developed in the 17th century by nearby Derbyshire beer breweries to fuel their malt kilns. The resulting coke was almost pure solid carbon and free of volatile compounds. As a fuel it had three major advantages – it was cheap and in plentiful supply, it burnt in an orderly manner, and it introduced far fewer impurities into the iron than had the burning of conventional charcoal. Darby successfully cast iron using coke in 1709. His main products were iron pots which were much cheaper and lighter than their predecessors. Even with these improvements, iron-making was a major consumer of coal with each ton of iron requiring about 8 ton of coal.

The coking technology spread slowly through the 18th century. Good chains became available due to the better iron-making and increasing demands of ocean-going ships for iron chains and anchors. The first European suspension bridge using iron chains as the main structural component was the Winch Bridge built in 1741 over the River Tee near Middleton in Durham.[161]

Further furnace improvements were required to produce useful large quantities of cast iron as a practical alternative to wrought iron. In 1766 the Darby family refined the furnaces which had previously been used for smelting bronze and copper. The process keeps the fuel separate from the ore and thus diminished impurities. However it also requires more heat and was only feasible for iron-making when coke became available. The product was a more malleable iron that was the structural equivalent of a wrought iron.

The pig and cast iron resulting from the 18th century processes contained about 4% carbon and the embrittling effect of carbon was discussed in the previous Chapter. The products also contained many non-iron impurities inherited from the iron ore and the fuel. Many of these exist as "foreign" *inclusions* trapped inside the iron crystal structure during cooling. The Shropshire ores were also high in sulphur, which can have further deleterious effects on iron properties. It was then found that adding limestone during the reduction process aided the removal of impurities. Many of the remaining impurity effects were relieved by wroughting and hammering.

In 1783 the ingenious and innovative Henry Cort,[162] working in the Lancashire – Shropshire region invented grooved rolling mills that permitted structural components to be produced from pig iron. His rolling mills were 15 times more productive than their predecessors. The resulting materials were still often called wrought iron, although the products were processed beyond the original wroughting and hammering, and often bypassed wroughting altogether. Cort produced a further major advance in 1784 with the introduction of a process known as *puddling*. This also copied bronze and copper smelting and involved stirring or shaking the molten iron in a reverberating crucible within the furnace. The process encouraged the chemical reduction of carbon oxides. The more malleable lower carbon iron was separated out as it formed as puddles in the higher carbon molten iron. Innovations like this evolved slowly as they were developed by empirical observation rather than by scientific analysis. Despite his remarkable work Cort died in poverty.[163]

In 1780 when the inventions of Darby and Cort were being unveiled in Britain the country was producing less iron than France. By 1806 the Merthyr iron works in Wales had become the largest in Britain, producing about 4000 ton of iron each year. In the 1820s the new iron-making developments had spread to Europe, beginning with the region around the coal fields near Liège in northern France and Charleroi in Belgium.

In the late 1820s James Neilson[164] in Scotland demonstrated the advantages of pre-heating the introduced air and the new steam engines described in Chapter 8 were used to drive pumps which increased the amount of air blasted into a furnace. As a further advancement, Neilson preheated the air using "waste" heat from the iron-making process.

Thus, as a result of steam-power, the quality and cost of iron had improved and it was available from a variety of sources in large quantities and in an increasing range of useful shapes. As a consequence iron production in Britain increased by a factor of ten between 1780 and 1830. By 1848 Britain – aided by larger and higher furnaces – was producing more iron than the combined output of the rest of the world.[165]

More broadly, the use of the steam engine for a whole range of diverse new purposes other than pumping water from mines – making iron, powering factory machines, hauling trains – is perhaps the best example of the "*knowledge spillover*" which some have used to explain the progress of the Industrial Revolution.[166]

CHAPTER 5C

TAKING ADVANTAGE OF THE IMPROVED PRODUCT

The beneficial effects of the improved iron-making processes snowballed and, by the end of the 18th century, cast iron was being used to produce bigger and better cannons for use in the Napoleonic Wars and for the cylinders and pistons of much-improved steam engines (Chapter 8). Indeed the two markets were closely linked as the technology developed to produce cannons with smooth internal surfaces, circular cross-sections and straight bores led directly to the manufacture of piston casings and shafts for the new piston engines. For example, James Watt's development of the condensing steam engine in the late 18th century is discussed in Chapter 8. In his initial work he had been greatly inhibited by the poor fit of pistons within their cylinders and once complained that the gap between a piston and a 700 mm diameter cylinder was "*the thickness of a little finger*".

The cannon situation had improved markedly in 1774 when Shropshire iron-maker, John Wilkinson,[167] developed a machine for boring a more precise internal cylinder in a cannon casting. Watt saw the value of Wilkinson's bored cylinders and they helped turn Watt's invention into a commercial reality. In a neat twist, Wilkinson then used the new engines to improve the output of the pumps blasting air into his iron-making furnaces. Known as "iron-mad" he was an iron enthusiast who actively promoted the use of iron and was proud of his iron boat, iron coffin and iron obelisk that he had made to mark his grave.[168]

The new steam engines required precision engineering to produce devices to control valves and to transmit power. Early in the 19th century Henry Maudslay[169] in London made major advances in machine tool technology, particularly in relation to lathes, drills, screws and cutting devices.[170] A major application of the new lathes was to provide precision threads on screws and bolts. Maudslay is rightly considered the father of precision engineering, working with accuracies as small as a few micrometres. As in other cases, the process was positively incestuous as the one of the important applications of steam power was to operate high speed lathes, drills and forges.

These machining applications required specialised steels and the next major step in this direction was in 1868 when Robert Mushet[171] added tungsten to steel to produce a suitable material for metal cutting tools. Mushet also discovered that adding manganese to steel made it better for forging. In 1887 Robert Hadfield[172] developed manganese steel which was harder and tougher than earlier steels in which increased hardness had been associated with increased brittleness.

Another consequence of the new irons and iron-working abilities was the development of better iron drill tips for geological exploration. One consequence of this was the discovery of many large petroleum deposits at mid-century (Chapter 15a).

By the 1830s iron was available in large quantities from a variety of sources. It was of improved quality and available in an increasing range of shapes. And very importantly, as a result of steam-power, the new metal products were better and cheaper and far more readily available than any of their predecessors. A fine illustration of the resulting consequences is provided by the dramatic development of 19th century bridges.

As noted in the previous chapter, iron had been used to make the links used for suspension bridge chains. In an appropriate coincidence, the larger scale structural use of cast iron began in Coalbrookdale which was the centre for the Darby family's generations of iron making. There, between 1777 and 1781, Abraham Darby[173] – a grandson of the Abraham discussed in the previous chapter – with the active support of John Wilkinson – constructed the first iron bridge, the strikingly graceful Iron Bridge over the Severn River (Fig. 23a & [174]). This was no small test bridge hidden from public view. Instead it was an enormous act of technical faith, meeting a strong demand for a local crossing of the Severn Gorge. The bridge was built entirely of cast iron (except for its stone abutments) and spanned about 30 m over a deep river valley. The design was largely due to Thomas Pritchard, a designer and builder from nearby Shrewsbury. Darby had insisted on the bridge structure being exclusively of iron and contributed much to the design and construction. The bridge used about four hundred tons of iron, cast in sand moulds formed on flat ground near the bridge site. The design details for connecting the cast iron pieces drew heavily on earlier jointing practice for wood. The bridge remains in use today, although vehicle use has been severely restricted since 1934.[175]

Fig. 23a The 1779 Iron Bridge over the Severn River at Coalbrookdale. The bridge height was chosen to accommodate the masts of the sail-powered cargo boats – Severn Trows –which used the river. Photo by the author.

Thomas Telford's fine career as a road builder was discussed in Chapter 4. In 1796 he built England's second iron bridge at Buildwas, also crossing the Severn, just 3 km upstream from Ironbridge (Fig. 23b). Telford carefully examined the structure used at Iron Bridge and consequently took more advantage of the compressive properties of cast iron. As a result, although his bridge had a longer (39 m) span and flatter (6 m) rise, it nevertheless used half the amount of cast iron that Iron Bridge had required. It survived until 1906.

Reducing the amount of material in a bridge had a range of advantages. Practically, it reduced the impedance that the bridge caused to flood flows in the valley. Aesthetically, it produced structures with lightness and some elegance (Figs 23 & 24). Telford's design principles of efficiency, economy and appearance within the landscape issued in a new engineering era. He argued that the primary visual purpose of a structure was to express its load-carrying function. Of the nine major cast-iron bridges built between 1779 and 1871, eight were due to Telford and five are still standing.[176]

Fig. 23b Telford's cast-iron bridge over the Severn at Buildwas (1796 – 1906). Drawn by Percival Skelton.[177]

A major technical advance occurred with his Craig-Ellachie Bridge in Moray in Scotland where Telford used Welsh cast-iron to produce an arch bridge with a 46 m span and a rise of only 6 m.[178] Completed in 1815, the bridge was refurbished in 1964 and remains in limited service today, carrying pedestrians and cyclists (Fig. 24). Note that the technical difficulty of a bridge increases with the square of the span and as the rise-to-span ratio decreases.

Fig. 24 Telford's 1815 cast-iron bridge over the Spey at Craig-Ellachie. via Wikipedia Commons.

A year later the Coalbrookdale Company provided the cast-iron arch Liffey Bridge (Fig. 25) with a slightly smaller span of 42 m but a more challenging span to rise ratio of 12.[179] Popularly known as the Ha'Penny Bridge, it too remains in use today. The main contractor was John Wilson from Coalbrookdale, who later became the contractor for the Menai Straits bridge discussed below. The longest of Telford's bridges in this genre was the 52 m span Mythe Bridge over the Severn at Tewkesbury (Fig. 26). Completed in 1826, it is still carrying traffic.

Fig. 25 The 1816 Liffey or Ha'Peny Bridge in Dublin, provided by the Coalbrookdale Company. via Wikipedia Commons (Pohl).

The move away from chains composed of cast iron links began in 1801 when James Finley[180] constructed a 21 m span suspension bridge over Jacobs Creek in Uniontown, Pennsylvania. He built his chains from 2 m long wrought iron bars looped around connecting pins. The bridge survived until 1833. Captain Samuel Brown[181] introduced wrought iron anchor chains into the British Navy in 1806 and in 1817 he patented the use of wrought iron eye-bars. He used this system in 1820 for the 135 m span Union suspension bridge over the Tweed River, near Horncliffe, between England and Scotland. The iron bar system was about twice as strong and half the weight of the traditional chains made of the contemporary iron links. The chains were replaced by cables in 1902.

Fig. 26 Telford's 52 m span Mythe Bridge built over the Severn at Tewkesbury in 1826. Photo by Philip Halling[182] via Creative & Wikimedia Commons.

Thomas Telford's 174 m span Menai Straits suspension bridge built in 1826 had a shipping clearance of 30 m (Fig. 27) and broke the span records previously held by the Union suspension bridge. Telford had been impressed by Brown's work and to exceed its record span he used cables composed of 2.8 m long wrought iron eye-bars produced by a nominated Shropshire ironworks. He carefully tested the bars to a load of twice the calculated in-service load.[183] The Menai bridge was a significant work and an impressive reminder of Telford's intuitive engineering skills, as it was designed by judgement and extensive testing, rather than by calculation (although by then the mathematics of the catenary had been solved). An important contemporary, physicist David Brewster, wrote that Telford "*had a singular distaste for mathematical studies, and never even made himself acquainted with the elements of geometry*".[184]

Fig. 27 Telford's 1826 Menai Straits suspension bridge in Wales. The important transport context of this bridge is discussed in Chapter 4c. See also Fig. 29. Early drawing by Percival Skelton.[185]

A famous American bridge builder, David Steinman, subsequently described the Menai Straits bridge as "*a bridge of noble vision and daring courage*".[186] The current bridge was rebuilt in 1940, using Telford's original masonry piers but employing steel cables rather than iron eye-bars.

Making wire cables by drawing iron though dies produced a similar beneficial effect to wroughting. Thus the more manageable and reliable wire cable began to replace eye-bar chains. The use of wire was also advanced by innovative work in Philadelphia in 1816 and it was then strongly promoted by Louis Vicat[187] in France.[188] The concept was applied by Marc Seguin[189] and Guillaume-Henri Dufour[190] who built small cable suspension bridges in France and Switzerland in the 1820s. Their first vehicular bridge was built in Geneva in 1822. The first wire cable bridge to exceed Telford's eye-bar record at Menai Straits was Joseph Chaley's[191] 273 m span *Gran Pont Suspendu* over the Sarine River at Fribourg in Switzerland (Fig. 28). Completed in 1834, it was demolished in 1923.

The bridges in Figs 24-29 are spectacular visual illustrations of the technical changes, skills and attitudes brought about by one of the first flowerings of the Industrial Revolution. They are in way reminiscent of the elegant coaches produced when iron became widely used in the coaching at the end of the 18th century (Chapter 1e).

Fig. 28 Chaley's 273 m span *Grand Pont Suspendu* over the Sarine at Fribourg, Switzerland. Completed in 1834. via Wikipedia Commons.

For applications involving stretching or bending, wrought iron would have been a better material to use but in the first half of the 19th century no method had been found to mechanise its production. In 1850 Britain had about 3000 "*puddling*" furnaces producing an average of less than 2 ton of wrought iron per day.[192] In addition, the structural properties of wrought iron can vary greatly as they depend on slag-type impurities and internal discontinuities in the finished product. Thus, there was a general need for a form of iron suitable for structural use. This need was to be satisfied by the development of mild steel in the second half of the century.

As discussed in Chapter 5a, steel had long been used as a "boutique" material for manufacturing fine knives and swords. The production of steel in sufficient quantities for it to be used for making machines and structures resulted from Henry Bessemer's[193] invention of the "*tilting convertor*" in 1850-6 at a Sheffield ironworks. His motivation was to produce better material for the armaments industry, in particular for the Crimean War. After many trials and modifications over some years he produced an egg-shaped device which blew large volumes of cold air through molten pig iron, allowing the oxygen in the air to chemically and exothermally combine with carbon and impurities in the iron, removing them as part of a gaseous carbon di-oxide effluent or as a scum-like slag. The chemical reaction was violent and spectacular.[194] The product was called *mild steel* and had better mechanical properties than wrought iron.

The major initial structural demand for mild steel came from the railway sector where the stronger and more ductile steel rails allowed the rail operators to expand their services. Trials of steel rails between 1857 and 1862 showed that although a steel rail was twice as expensive, its service life was up to sixteen times greater than that of a wrought iron rail.[195] The armaments industry also became a major user of the new steels. The use of steel plates also greatly improved the construction and operation of high pressure boilers needed for steam engines.

In 1866 higher furnace temperatures were being achieved by using an open hearth process developed in England by Carl Siemens.[196] This method allowed greater tonnages of better quality steel to be produced. Another major increment occurred in 1877 when the cousins Percy Gilchrist[197] and Sidney Thomas[198] enhanced the Bessemer method by inventing the *basic* steel-making process at an ironworks in Blaenavon, Wales. Their discovery was that replacing acidic linings of the bowl holding the molten iron by a *basic* lining such as limestone would absorb many of the impurities (particularly phosphorus oxides) in the iron ore. This not only made better steel, but also allowed a wider range of iron ores to be used.

Later in the 19th century the greater availability of power and of lengths of iron made rolling increasingly more feasible. The resulting rolled products were immediately useful. In addition, the rolling enhanced the properties of the steel in a similar manner to the gains that had been reached by the earlier wroughting processes.

The comparison was not only between the brittle cast irons and the ductile wrought irons, as the structural advantages of the geometric shape of rolled sections – such as rolled steel joists – were enormous. In terms of strength and stiffness, the rolled section was about twice as cost-effective as the equivalent solid bar. Furthermore, the sections could be readily connected by riveting (and later bolting) though their relatively thin flanges and webs.

Once mild steel became available, it was soon used in a vast range of industries – locomotives, railways, bridges, ships, boilers, chemical manufacture, agriculture, machine tools, and construction.[199] Many of these energies and applications are explored elsewhere in this book. It has been said that the production of mild steel "*unleashed the full energies of the Industrial Revolution throughout the world*".[200]

CHAPTER 5D

USING AND UNDERSTANDING IRON AND STEEL

The Holyhead railway was a strategic rail link which was discussed in Chapter 4c and will be examined further in Chapter 11a. It included two challenging major bridges – the Conwy Bridge over the River Conwy and the Britannia Bridge over the Menai Strait. They were both innovative designs produced by Robert Stephenson,[201] son of George Stephenson the famous railway engineer whose career is discussed in Chapter 9d. The trains passed through large rectangular tubes constructed from pieces of wrought iron plate riveted together and stiffened with rolled sections in a manner that had been developed by the builders of iron ships (Chapter 9f &[202]). A cellular top flange was used to enhance resistance to lateral buckling.

The bridges both opened in 1849 and clearly demonstrated the effective use of wrought and rolled iron and of structural rivets. Riveting dates from the first use of iron in tools and weapons but it had come into prominence in about 1810 when the new steam engines discussed in Chapter 9b created a need for high-pressure boilers and well-sealed cylinders.

The design move from the suspension and arch forms used by Telford and others to a large box working as a beam was made because the suspension form was thought to be too flexible for railway loadings and that the arch would not provide adequate shipping clearance beneath the bridge. Originally the box was to be partly supported by suspension cables but it was soon realised that it would have adequate stiffness in its own right. The Britannia's pair of tubes were continuous over four spans of 72, 142, 142 and 72 m and were about 8 m deep (Fig. 29). The Britannia Bridge was still in service when it was destroyed by fire in 1970. The Conwy survives but is now only used for pedestrian traffic.

The detailed engineering testing and design behind the two box bridges was done by William Fairbairn[203] and Eaton Hodgkinson.[204] Before the testing it was thought that the critical feature would be tension cracking of the bottom flange of the box as a result of bending stresses. However the testing alerted the designers to a potential problem with plate buckling of the top flange. This was the first study of thin-walled structures failing due to instability.[205] Further work on instability and buckling is discussed in Chapter 7a.

Fig. 29 Robert Stephenson's wrought iron Britannia Bridge over the Menai Straights. Telford's earlier road bridge (see Fig. 27) can be seen in the background. Drawn c1850 by S. Manning and S Green. Wikipedia Commons

Despite the Britannia Bridge and the many spectacular structural successes discussed in the previous Chapter, 19th century engineers were much less successful with using cast iron in a wider range of circumstance and there were many 19th century bridge failures. The failures had three main common causes and two of these, brittle fracture and fatigue, related directly to the material.

The basic problem was that the behaviour of iron under load was not well understood. Cast iron, like masonry, is very effective in carrying compressive stresses but has an inherently low tensile strength. This tensile strength depends on subtleties in the internal structure of the iron. Cleavage cracking of iron and steel when subjected to tensile forces can occur under a single application of a load if the stresses are high and/or the temperatures low. This is commonly called *brittle* fracture. The stresses may be directly caused by the load but are often due to stress concentrations around internal defects in the iron or introduced during fabrication. *Fatigue* cracking occurs when successive passages of a load cause a small crack to grow incrementally until it reaches a critical stage.

At the time when iron and steel usage was burgeoning, both these processes were poorly understood.

The third failure cause related to the application of the material. As will be discussed in Chapter 10, the same steam power that permitted the production of iron in large quantities also led to new railways powered by large steam locomotives. The Stephensons' Rocket weighed 4 t and locomotive weights would soon exceed 25 t. When combined with a string of fully-laden wagons, the new railways applied loads to bridges which were of a size and frequency far beyond that which had been experienced by contemporary road bridges.

The consequence of the presence of these three potential failure causes was dramatically illustrated by a number of failures of cast iron bridges. Perhaps the best known was the failure in May 1847 of the bridge carrying the new Liverpool-Holyhead railway (see Chapter 11a) over the River Dee near Chester in England. The bridge had been designed by Robert Stephenson and had been in service for less than a year. It had three 33 m spans for which Stephenson used simply supported cast-iron beams. Aware of the brittleness of cast iron, he employed draped wrought-iron bars to carry the tensile forces. The method had first been used by Charles Vignoles in 1831.[206] Unfortunately, the methods of analysis then in use did not show that there would still be tensile stresses in parts of the cast iron beams and these failed by cracking during the passage of a train, killing five passengers. Whilst the precise cause of the failure is still subject to some debate,[207] it properly marked the end of the use of cast iron beams and hurried their eventual replacement by steel beams. Many of the other failures of cast iron bridges were due to metal fatigue and these events will be discussed below.[208]

A broader lesson from the Dee failure was that the bridge was designed and built in an era of great confidence and enthusiasm, particularly in time regarded as a period of "*railway mania*" (see Chapter 11b). Petroski[209] described Stephenson's error as due to a "*success syndrome*" which allowed the material and the design to be taken "*to the outer limits of experience.*"

The problem was circumvented by moving to wrought iron and then to steel, although even these materials needed to be treated with respect to avoid brittle fracture.[210] A relevant case was the move from chains of cast iron links to wrought iron eye-bars to wrought iron cables discussed in the previous Chapter.

Before the Revolution, bridge and building design had been very much a specialist trade in which masons built following ancient and well-tested geometric rules as in Vetruvius' 1st century BC building manual, *De Architectura*.

An initial break-through occurred in 1676 when Robert Hooke proposed the core concept of material elasticity. When a piece of iron is loaded, in the absence of brittle behaviour, it deforms in proportion to the load applied. When that load is removed, the iron piece returns to precisely its original shape. This is called *elastic* behaviour. Eventually a load is reached at which permanent deformation occurs. This is called *yielding*. There was an obvious strong operational incentive to keep stresses well below the yield level. This was complicated by a growing appreciation that steel structures could contain stresses produced in the past – e.g. during manufacture – and that these "*residual*" stresses could lead to yielding in service at lower than expected loads.

Earlier structural materials such as wood and masonry produced structures which were relatively massive. But, as demonstrated splendidly by the new springs and frames for coaches (Chapter 1f) and the bridges described in the previous Chapter, iron can be used in more slender proportions. This exciting development produced a new set of issues when iron was used in compression. European practice used the work[211] of the mathematician Leonhard Euler[212] in Switzerland in the 1740s and based design on the elastic buckling of a straight column (Chapter 7a). Indeed, in the 18th century some of the world's best mathematicians had worked in Europe on the theory of structures.[213] The process was greatly aided in 1747 when France established the École Nationale des Ponts et Chaussées to train bridge and road engineers (Chapter 3b). Famous mathematically-oriented associates of the school included Cauchy, Navier and Fresnel. This initiative will be pursued further in Chapter 7 which examines the development of engineering theory.

Iron was pushed further towards its application to structures in 1807 when Thomas Young published his "*Course of lectures on natural philosophy and the mechanical arts*".[214] However, in the 1820s iron came to the fore as a sophisticated material when Thomas Tredgold[215] took Young's work and produced a well-used text book[216] on the strength of cast iron. Tredgold had trained as a carpenter and then became an architect. This led him to the testing of wood and cast iron to determine their structural properties.

The availability and wide use of steel also provided a further incentive for better understanding of its behaviour.[217] The careful "scientific" study of steel began in the 1860s. In 1863 Henry Sorby,[218] an amateur geologist, polished and acid-etched the surfaces of steel specimens. This allowed him to observe the distinctly non-uniform internal structure (grain boundaries, laminations, etc) of the steels. In the same period David Kirkaldy[219] tested steel specimens to destruction in his commercial testing laboratories in London. The first measurements of the behaviour of iron beyond the elastic range were made by August Wöhler[220] in the 1860s.[221] In the 1870s Johann Bauschinger[222] began his extensive studies of the yielding of steel. In the 1890s Friedrich Engesser[223] applied the yielding concepts discussed above to the buckling of columns (Chapter 7a). By this time, mild steel had been in wide use for half a century.

In the 1830s the French were describing the increasingly common failure of coach springs as due to *fatigue*.[224] The problem was exacerbated as the new steam engines and steam railways began to cause high repetitions of load on a much larger scale than ever produced by road coaches. For example, in 1842 some 55 people were killed in a train crash due to an in-service axle failure on the Paris-Versailles line. The first formal report on *fatigue* appears to have been in an 1843 lecture by Macquorn Rankine[225] on the failure of axles. The issue was pursued further by Frederick Braithwaite.[226] Similar railway axle failures in service were occurring in Germany and this led to August Wöhler developing the now-common experimental method for the fatigue testing of critical steel components.[227]

Nevertheless, in 1860 two cast iron beam bridges were to fail under traffic – at Wootton in Warwickshire and the Bull Bridge in Derbyshire. The problem was more widespread than these English examples. Between 1870 and 1890 twenty-five percent of all iron or steel bridges in the USA – 520 of them – failed due to metal failure and/or poor design.[228]

A notorious and relevant failure example was the collapse in 1879 of the Tay Bridge in Scotland after the truss bridge had been in service for only 18 months. Seventy five people were killed. The bridge was built from both cast iron and wrought iron and was then the longest in the world. In retrospect, both the designer and his key advisor, the Astronomer-General, had made quite unrealistic and implausibly low assumptions about wind pressure,[229] although other factors such as the fatigue of cast iron lugs probably also contributed to the failure.[230] Three years earlier, the fatigue failure of a cast iron lug had been the prime cause of the collapse of the Ashtabula Railway Bridge near Cleveland, Ohio which led to the death of ninety two people.[231]

Thus the lessons about fatigue were being learnt in a hard and tragic way. Steels were also often failing in a brittle fashion although steels that are well-made are not brittle. The prime need was for good steel-making and then for a suitable test of steel quality.[232] The first test to attempt to detect steels prone to brittle fracture was the Charpy Test which was only introduced in 1901. The crack resistance of a steel comes from it yielding rather than cracking when subjected locally high stresses. Thus, as well as a need for better steel-making, there was a need to avoid causes of high local stresses, such as sharp changes in the geometry of steel member. Such sharp changes mean that even a modest external load will produce a local stress able to cause a local crack. Each repetition of that load will cause the crack to grow until it is of sufficient size to cause failure of the object. This was the cause of many of the early axle failures. The lesson was slowly learnt.

Observing the resulting fracture surfaces by eye or by low-powered microscopes caused people in the 19th century to believe that the metal had changed from "*fibrous*" to "*crystallised*" or that there had been "*molecular disorder*". These misinterpretations confused investigators for the most of the 19th century – indeed, the misinterpretation was repeated as late as 1979.[233] What investigators were actually seeing with their low-powered optics was not *crystallisation* but the optical facets of the fracture surfaces. At the best it was a curious interpretation as iron is inherently crystalline with strong atomic bonds between iron atoms. Similarly, the pre-fracture "*fibrous*" aspect of the misinterpretation stemmed from the internal patterns produced by the wroughting processes. Interestingly, as early as 1850 Robert Stephenson had viewed fatigue fractures with a relatively powerful microscope and seen no sign of crystallisation, leading him to caution against the crystallisation theory.[234]

The first steel bridges copied wrought iron practice. These were three Dutch railway bridges built in 1863 and a railway swing bridge built in 1864 near Warrington in Lancashire. The first major bridge to make extensive use of mild (structural) steel was built in 1874 by James Eads[235] over the Mississippi River at St Louis. It has three truss-arch spans of 151 m, 156 m and 151 m using tubular steel members for its chords. The bridge used a mixture of steel and wrought iron and Eads had trouble obtaining suitable steel. Nevertheless, the bridge remains in service today (Fig. 30). The first all-steel bridge was built in 1876 over the Missouri River in Montana. Rolled wrought iron beams had been available since the 1850s. The first rolled (mild) steel beams were produced by Dorman Long in Britain in 1885. The new beams revolutionised the design and construction of short-span bridges.

Many have described the railway bridge crossing over the Firth of Forth between Edinburgh and Fife in Scotland as the pinnacle of Victorian engineering (Fig. 31). The bridge was designed by John Fowler[236] and Benjamin Baker[237]. Built between 1882 and 1890 it has a total length of 2.5 km. Its various new features were brutally apparent. A later famous bridge builder, David Steinman, commented that the "*Forth Bridge seems to scorn beauty, even seems to revel in a certain angularity, for it boasts of invincible strength, in which it glories with superb indifference and insolent pride.*"[238] Perhaps a fitting structural engineering end to the 19th century?

Fig. 30 Eads' 1874 St Louis steel bridge over the Mississippi. by Mitchell Schultheis via Wikipedia and Creative Commons.

Fig. 31 Fowler and Baker's 1890 Firth of Forth railway bridge. Photo by the author.

The truss bridge form had been largely developed in the USA where wood supplies were plentiful. It soon became apparent that the wooden tension members in the trusses could be more than usefully replaced by wrought iron and later by steel ties. Thus trusses had changed from wood to steel during the course of the 19th century bringing all the advantages of iron and steel and all the above problems associated with the material's inappropriate use.[239]

The reinvention of cement in the 1820s is discussed in Chapter 17c. Concrete is a mixture of cement, water, broken stone and sand and was reintroduced at the same time. The major new construction material of the 19th century was *reinforced concrete* which combined concrete and its high compressive strength with the relatively cheap mild steel rods and their high tensile strength. The technique had been used in a minor way since Roman times[240] and in 1824 Marc Brunel[241] used iron bars to improve the tensile strength (reinforce) the concrete and brick components of his tunnels.

The widespread availability of mild steel increased interest in using it to enhance the strength and reduce the cracking of concrete castings. Parisian gardener Joseph Monier[242] patented one such system, firstly in 1868, and for the next decade or so he patented a range of applications. He placed the reinforcing at the centroid where it only controlled thermal and shrinkage cracking. Working in London in the 1870s, an American – Thaddeus Hyatt[243] – demonstrated the strong bond that formed between concrete and iron – aided by their very similar coefficients of thermal expansion.[244] His 1878 patent showed his structural perception as it also specifically stated that the iron should be placed "*on the tension side*" to raise the bending strength of the cross-section.[245] Gustave Wayss[246] then applied Monier and Hyatt's development to a range of applications in Berlin.[247] The first significant reinforced concrete structure was built in 1899 when a mason, François Hennebique,[248] constructed a triple-arch (40 m, 52 m, 40 m spans) reinforced concrete bridge over the la Vienne River in Châtellerault in France.

In the course of the 19th century, iron and steel production increased 30-fold to 9 million ton. As many examples in this book show, during that time iron had created and then serviced vast new markets. Meanwhile, the countryside was not unaffected by the changes occurring, as the next Chapter will illustrate.

PART 6

THE BLACK COUNTRY

CHAPTER 6

THE BLACK COUNTRY

England was not a technically advanced part of Europe in the period before the Industrial Revolution. Indeed, throughout the Middle Ages it was relatively backward and was a source of raw materials rather than of manufactured products.[249] The 17th century flowering of intellectual giants like Isaac Newton, Robert Hooke and Robert Boyle was an early harbinger of change.

The history of small scale iron-making was explored in the Chapter 5a and was seen to have occurred in many countries throughout the world. The Domesday Book records that Gloucester was a centre of the 12th century English iron trade. There were micro-works in the Forest of Dean and further up the Severn valley at Clee Hills and Wombridge near Coalbrookdale in Shropshire. Indeed, iron-making was common in southern Britain, from Anderida near Crawley in Sussex to the Aberdare / Merthyr Tydfil region of Wales. These small iron works commonly used wood for fuel, however the economics of larger-scale iron-making was to depend on the ready availability of coal.

Chapter 5c then showed how many of the seeds of the Industrial Revolution germinated in Coalbrookdale and its hinterland. The reasons for this more specific change are predominantly that, as well as accessible iron ore deposits of high purity, there were also:

* large and accessible coal seams about 10 m thick south-east of Coalbrookdale (the coal had a low sulphur content which led to good iron making, Chapter 5b),
* limestone quarries at Benthall a few kilometres to the south,
* good transport for heavy products using boats on the River Severn (Chapter 5c), and
* men who were skilled in operating ovens and furnaces.

For example, there was a pig iron foundry at Horsehay 10 km north and an ironworks at Willey, about 10 km south of Coalbrookdale. There had also been iron-making at Coalbrookdale itself from at least the 17th century. A 1658 blast furnace can still be seen today at The Ironbridge Gorge Museum. Abraham Darby changed the scene early in the 18th century when, in the Coalbrookdale region, he began to smelt iron ore using coke from coal, rather than using wood. The benefits of using coke rather than coal were discussed in Chapter 5b and the competitive rivalry between the fuel suppliers is nicely illustrated in Alken's 1825 cartoon (Fig. 32).

Fig. 32 Alken's view of competition between coal and coke in the new steam age. Henry Alken was a prolific producer of paintings and drawings between 1815 and 1830.

The well-known commentator, Arthur Young, has already been encountered in Chapter 4b. When he visited Coalbrookdale in 1785 he described it as "*a very romantic spot*" but, in a portent of the future, he also said that the countryside was:

> "*too beautiful to be much in union with the variety of horrors spread at the bottom; the noises of forges, mills, with their vast machinery, the flames bursting from the furnaces with the burning of coal and the smoke of the lime kilns*".[250]

Loud explosions also continually emanated from the various iron-works.[251] The scene in 1801 is captured in de Loutherbourg's contemporary painting (Fig. 33).

Fig. 33 Philippe Jacques de Loutherbourg's 1801 painting *'A View of Colebrook Dale by night'* featuring night work at the furnaces. Compare this with Fig. 23a, where nature is seen to have recovered some territory. With permission of the Science Museum, London (10307414).

The smelting of iron ore from coke was first used, albeit unsuccessfully, by one of Darby's ancestors, the metallurgist Dud Dudley, who had lived in Dudley in the 17th century. Dudley is about 30 km south-east of Coalbrookdale. The large Conygree colliery was on the eastern edge of the old town. In a circuit typical of the times, it will be seen in Chapter 9 that Darby's cast iron was used by Newcomen at his Dudley plant to make the steam engines which were used to pump water from the nearby Conygree colliery, where one of the first of his pumps was installed in 1712. Its cylinder had a bore of about 400 mm and a length of over 2 m. Another was installed at the same time in Wolverhampton. Similarly, Newcomen's engines were also used in place of bellows to dramatically improve the air blast in Shropshire iron-making blast furnaces.

Consequently, Dudley became the centre of a region known as the Black Country, a term particularly applied to the area between Birmingham and Wolverhampton, but often extended south via Halesowen over the Staffordshire coalfields and west through Wolverhampton to

Coalbrookdale. The region was called the Black Country during the Industrial Revolution as the smoke, grime and soot from all the furnaces and ovens and steam engines and factories – those *dark satanic mills* – had literally turned both the sky and the ground black. In a famous story, when Queen Victoria passed through the region in the Royal Train she ordered the blinds of her carriage to be pulled down so that she did not have to endure the scene. The area, including the ruins of the old Dudley castle, is now part of the Wren's Nest National Geological Reserve. Furthermore, an American observer in the 1860s commented that "*the Black Country is beautifully framed by a Green Borderland and that border is rich and redolent with two beautiful (natural) wealths*."[252]

The words *dark satanic mills* are from a poem by William Blake in about 1804 and which is now best known as the words of the anthem *Jerusalem*. The timing is very early in the Industrial Revolution and it is thought that Blake, at one level, may have been referring to a contentious flour mill near his home in Southwark, which had been the first major factory in London. It had used a Boulton and Watt steam engine (Chapter 9) to grind wheat into flour. Another major factory application was the use of Edmund Cartwright's[253] automatic powered loom for weaving cloth, invented in 1784.

Blake's views were certainly in marked contrast to those of Erasmus Darwin[254] who, a decade earlier, had written his poem *The Botanic Garden*, which had:[255]

> *celebrated the new engines and machinery transforming the age and had described the steam engine as a monument to both motion and order with its piston "resistless", sliding through its iron walls.*

The actual poem is presented in the next Chapter.

James Nasmyth[256] was a significant contributor to the Industrial Revolution, having created a major machine tool industry and in 1840 inventing the steam hammer used to forge large pieces of metal into useful shapes. In September 1830 he travelled through the Black Country and his notes (published posthumously) portray the countryside at its worst:[257]

> *I then proceeded to Dudley. The Black Country is anything but picturesque. The earth seems to have been turned inside out. Its entrails are strewn about; nearly the entire surface of the ground is covered with cinder-heaps and mounds of scoriae. The coal which has been drawn from below ground is blazing on the surface.*
>
> *The district is crowded with iron furnaces, puddling furnaces, and coal-pit engine furnaces. By day and by night the country is glowing with fire, and*

the smoke of the ironworks hovers over it. There is a rumbling and clanking of iron forges and rolling mills. Workmen covered with smut, and with fierce white eyes, are seen moving about amongst the glowing iron and the dull thud of forge-hammers.

Amidst these flaming, smoky, clanging works, I beheld the remains of what had once been happy farmhouses, now ruined and deserted. The ground underneath them had sunk by the working out of the coal, and they were falling to pieces. They had in former times been surrounded by clumps of trees; but only the skeletons of them remained, dead, black, and leafless. The grass had been parched and killed by the vapours of sulphurous acid thrown out by the chimneys; and every herbaceous object was of a ghastly grey – the emblem of vegetable death in its saddest aspect.

Vulcan had driven out Ceres. In some places I heard a sort of chirruping sound, as of some forlorn bird haunting the ruins of the old farmsteads. But no! the chirrup was a vile delusion. It proceeded from the shrill creaking of the coal-winding chains, which were placed in small tunnels beneath the hedgeless road.

The grim and grimy scenes described above were soon replicated in Pittsburgh in the USA, Lille in France and the Ruhr in Germany. Throughout the industrialised world such devastated landscapes – shocking reminders of times past – were to linger and last for over a century. Nevertheless, such gloomy views must be seen in the context that the relatively small regions encompassed by the Black Country and its kin were many orders of magnitude more productive than any previous human endeavours in creating wealth and prosperity for the whole wide world.

The stories of the myriad of workers and their families who were integral to and yet unrecognised participants in this and other parts of the Revolution are not central to the theme of this book and so will not be explored further here. There are good accounts of the lives of ordinary folk during the 19th century; in particular Briggs 1987, Burnett 1974, Clark 1972, Kitson Clark 1962, Thompson 1963 and Wise 2008.

PART 7

TECHNICAL EDUCATION

CHAPTER 7A

THE BEHAVIOUR OF IRON AND STEEL COLUMNS AS A CASE STUDY

The story of iron in Chapter 5 demonstrated how the Industrial Revolution created a new class of creative practitioners. It became evident that experience and intuition and passed-down empiricisms would be insufficient if further progress was to occur. The new practitioners needed to not only know how to make things happen – they also needed to understand why those things happened and how better outcomes might be achieved by thought and planning rather than by trial and error. In this context the practice of *engineering* arose to utilise the insights of mathematics and science to create many new and wondrous worlds. Nevertheless, the merging of empirical knowledge and theoretical knowledge was slow and stumbling.

For example, elastic behaviour was defined in Chapter 5d. Elasticity and geometry determine the *stiffness* of a structural component. Determining *strength* is another matter and was usually empirically based. Bending and tension tests for establishing both the strength and stiffness of structural members were well established, particularly via the 18th century work of Pieter van Musschenbroek in Leiden in Holland.[258] However simple theoretical assumptions of elastic behaviour were certainly not appropriate when the British began producing large quantities of high compressive strength cast iron and high tensile strength wrought iron and started using them very visibly, as with Telford's suspension bridge which spanned 176 m and used wrought iron eye-bars to form the chain (Fig. 27). As noted in Chapter 5c, Telford carefully tested all the eye bars to be used in the bridge. There was no reliance on elastic stress predictions. The daring and visually impressive bridge was built between 1819 and 1826.

An ironical but inadvertent reversal of the empirical approach was due to Charles-Louis Navier[259], then the main French theoretician working on suspension bridge design. He had published a book[260] on the subject in 1823 and had visited England twice (1821 & 1823) to observe English work. He then designed a 167 m suspension bridge over the Seine in Paris which had to be dismantled in 1826 when partly constructed, due to inadequate horizontal anchorage of the cables.[261]

Designing for loads in compression proved quite challenging. A specialised but relatively simple example of this is in the design of iron and steel columns. Buckling and instability are related physical phenomenon which became practically relevant with the availability of iron plates. The phenomena are easily understood by considering the analogy of a smooth ball in a smooth bowl placed on a table. Disturb the ball a little with your finger and it will soon return to the bottom of the bowl. The situation is *stable* – disturbances have only a short-lived effect with no long-term consequences. However, if the bowl is flattened until it is a smooth flat plate, any disturbance of the ball will see it roll off the plate and the table and drop on the floor. Even a small disturbance will have a relatively large effect which will not be reversed after the disturbance has ceased. The ball is now on the floor. The situation before you disturbed the stationary ball ever so gently was *unstable*. The same situation occurs when columns are loaded in compression. Eventually as the load is increased or the column is made more slender, it becomes unstable and buckles. The buckling load thus depends on the stiffness of the material, but not on its strength.

Chapter 5d mentioned that European practice for the design of cast iron columns came to be based on the work of the mathematician Leonhard Euler in Switzerland who had investigated the elastic buckling of a straight column.[262] Euler's 1744 "*elastic line*" theory of the loading of columns produced the equation still used today for predicting the elastic buckling load of a column. However, columns slender enough to buckle elastically are too flexible for most building applications and so elastic buckling has rarely occurred in practice. Nevertheless, elastic behaviour produced neat formulas for column load capacity and made it possible to apply differential and integral calculus to structures. Consequently, assumptions of elastic behaviour prevailed in France throughout the 19th century.[263] The difference between British and French approaches had been noted in 1813 by Michael Faraday (Chapter 15) who had observed during his first visit to France that the French "*reason theoretically, without demonstrating experimentally, and errors are the result.*"[264]

The work of Eaton Hodgkinson and William Fairbairn in testing the components for Robert Stephenson's revolutionary box girder bridges was discussed in Chapter 5c. Hodgkinson was a farm boy who was encouraged by workmates to teach himself mathematics and science and was later tutored by John Dalton, the scientist who introduced the atomic theory of materials. Fairbairn had completed an engineering apprenticeship at a colliery and subsequently opened his own engineering business. In the late 1830s Hodgkinson conducted compression tests on cast iron columns in Fairbairn's foundry and his account of the work was published by the Royal Society in 1840.[265] The practical imperatives associated with the work meant that when Hodgkinson tested cast iron (and wooden) columns a century after Euler's work, many were of practically realistic proportions and did not fail elastically.[266] The short ones failed by crushing of the iron and the less slender ones by cracking caused by bending of the column due to eccentric loading or crookedness of the column.

Hodgkinson's tests were of high experimental quality. However, consistent with one of the themes of this Chapter, he made no attempt to explain his results in mechanistic terms by comparing failure modes to geometric and material properties. Indeed, in stark contrast to Faraday's observations of French attitudes quoted above, it has been properly observed that "*Anglo-Saxon researchers initially seemed reluctant to examine the scientific principles behind the physical properties of materials*".[267] Belatedly, Macquorn Rankine in the 1860s noted that reality lay somewhere between Euler's theory and Hodgkinson's empirical tests.[268] In this and many other areas of engineering, Rankine was often to be the voice of reason opposed to the empiricisms of his confrères.[269]

In England the design of columns was based on the work of Tredgold who ignored buckling and assumed that real columns were initially bent and calculated stresses in them by assuming an initial crookedness.[270] Hodgkinson's tests were later used by Lewis Gordon[271] to assign values to the empirical terms in Tredgold's by then widely-used formula for column design.[272] This produced an empirical formula for column strength with the empirical terms covering eccentricity and crookedness. It remained in widespread use for over a century.

Thus the work of Hodgkinson and Fairbairn formed the basis of iron and steel column design in Britain and many other countries for at least the next 150 years, despite the marked changes in fabrication techniques and material properties during that time. In the 1860s Rankine in Glasgow showed how the theoretical French approach could be rationally combined with the empirical British practice.[273] However, the difference between the straight, concentrically loaded, pin-ended elastic column of Euler's theory and real columns was instead covered by a variety of empirical adjustments. Some favoured approaches assumed that the real column was perfectly straight, but eccentrically loaded – these were called *secant formulas* as the loaded column adopted a secant shape; others assumed a Fourier transform for the initial (crooked) shape. The main virtues of the most favoured methods were that their algebra was neat and led to tidy "solutions." The final form widely in use at the end of the 20th century was the Perry-Robertson formula, which was still based on the early Tredgold-Gordon method and still unrealistically assumed that compression failure happened when yielding first occurred in a concentrically loaded column free of residual stresses, and with empirical levels of crookedness.[274]

One of the curious aspects of this long saga of empirical formulas for column design is that they were over-influenced by the behaviour of slender columns. Although mathematically and physically intriguing, most practical columns are relatively stocky and would fail by squashing and localised buckling of the plate elements that make up the cross-section, and would be greatly influenced by the bending stiffness of the members bringing the load to the column. The intricate algebra of a century of column formulas described above was largely unnecessary.[275]

The infrequent use of the relatively impractical slender columns has meant that there were relatively few failures due to buckling of slender members, although a number of steel truss bridges in Russia and Western Europe did fail when the top, compression, chord buckled laterally due to non-existent or inadequate lateral bracing of the top chord.[276]

Another more ancient example of empiricism in structural engineering was the variety of purely geometric methods used for arch design, first developed and applied by the early masons and famously documented by Marcus Vitruvius in his De Architectura in the first century BC.[277]

Bridges certainly provide some useful examples of the processes operating during the 19th century. Chapter 5 described how early in the century Thomas Telford in Britain used his great engineering intuition – and very little structural analysis – to produce a new generation of elegant and effective arch bridges by using cast iron in compression (Figs 25 to 29).[278] A decade or so later, by the testing and trials discussed above, Fairbairn and Hodgkinson were able to produce suitable structural members for Robert Stephenson's massive but effective Britannia iron box girder bridge described in Chapter 5d (Fig. 30).

However, when iron and steel became cheaper and more widely available, a different set of skills was required. Analysis became more critical than intuition. Thus, between 1884 and 1889 Gustave Eiffel[279] and his colleagues were able to analyse, design and build a 324 m high wrought-iron tower in Paris which now carries his name and was the world's tallest building until 1930. The Eiffel team had previously produced a spectacular wrought-iron bridge – the 160 m Maria Pia bridge over the Douro River in Porto, Portugal (Fig. 34). At the time it was the world's longest arch bridge. Eiffel had won the contract for the bridge with a price that was 31% cheaper than the closest alternative bid. The form of the Eiffel Tower was first suggested in the piers of four railway viaducts which he had built in 1867-9 for a railway track between Gannat and Commentry in France. Eiffel had trained at the École Centrale des Arts et Manufactures, leading French engineering school in Paris.

Fig. 34 Eiffel's 1876 Maria Pia bridge over the Douro River in Porto, Portugal. Photo by Joseolgen, Wikipedia Commons.

The iron bridges of Telford and the iron structures of Eiffel were bookends for the 19th century although the events described above were reflected in many other aspects of 19th century technology. These reflections will become apparent in following Chapters which discuss the development of useful steam engines and the invention and application of internal-combustion engines, which in most ways were even greater bookends than the iron bridges.

The Revolution had taken all technologies to a stage where empiricism was no longer the sole basis for further progress. There also had to be a major application of fundamental principles of science and engineering. But both the French experience with a concentration on scientific principles and the British reliance on empirical testing showed that final success only occurs when the two approaches go hand in hand.

CHAPTER 7B

TECHNOLOGICAL DIFFERENCES BETWEEN BRITAIN AND CONTINENTAL EUROPE

The 17th and 18th centuries are often seen as the Age of Reason, when the human condition was advanced by the application of science and by putting aside long-held views that had been supported by tradition and religion rather than by rational thought. In the new Age, conclusions were to be based on logic and evidence, rather than on superstition and inherited precepts. This change was only slowly working its way through society and its impacts varied widely. It occurred most rapidly in Britain where attitudes were relatively liberal and free from major religious restraints.

As noted in earlier Chapters, the 18th and early 19th centuries had produced some extraordinary people who caused major technical advances, working alone or in small groups. The column discussion in the previous Chapter usefully introduced a key difference that had emerged between British and European practice during this time. The British entrepreneurs and engineers often had only minimal formal education. For example, two of the greatest engineers of the 19th century were Thomas Telford and George Stephenson. Telford's education could, at best, be described as minimal (Chapter 4a) and Stephenson had had no formal education or training and, at 17, was still illiterate (Chapter 9b). Educated British technologists would have been schooled in the classics and then would have worked in industry and been strongly driven by its profit motive. The new breed of engineers would have gained their professional credentials by working as a "pupil" (or apprentice) with a recognised engineer. This approach worked well when the relevant technology was a simple and major part of the system in which it operated.

The story of an otherwise anonymous Thomas Wood is revealing. Born in Yorkshire in 1822 and described as "an intelligent boy", from the ages 6 to 8 he was sent to Grammar School where he was taught only Latin grammar (and no arithmetic) for two years before going to work in a woollen mill. At 14 he was apprenticed to a local mechanic. In his 20s, his career peaked when for a short time he was working with the latest Whitworth machine tools and gauges.[280]

Continental Europe, on the other hand, had a stronger system of technical education and a greater ability to reward people for their technical contributions alone. To become an engineer first required undertaking and passing detailed courses in science and mathematics, as well as engineering. Indeed, in the system developed in France late in the 18th century, French engineers would do a three-year core course at the École Polytechnique and then move to places like the École Nationale des Ponts et Chaussées for their engineering studies. The École Nationale des Ponts et Chaussées had been established in 1747 and the broader École Polytechnique opened in 1794 (Chapter 3b), half a century ahead of similar developments in England.

Germany had been heavily influenced by France but in the post-Napoleonic era, it changed the French pattern and introduced engineering schools that covered science and mathematics in the first two years and then engineering in last two years. Another difference was that the French undergraduates were mainly prepared for government or military service, whereas the German schools were focussed on industrial careers and worked closely with industry. As Chapters 14b&c will demonstrate when the careers of men like Otto, Daimler, Maybach and Benz are described, the German approach was to prove very fruitful in the second half of the 19th century. For example, the Karlsruhe Polytechnic Institute became a leader in teaching machine design.[281] Chapter 16 will show that in the second half of the 19th century many innovative French engineers were also working in the private sector.

Whilst the social and educational paths taken by young engineers and inventors in France and Germany was clearly important[282] it is very relevant to note from the examples already given and by the careers to be described in Chapters 14b&c that relevant employment during and after training had a major impact on the productive output of talented people.

On the other hand, Bronowski comments that in 19th century England "*the two universities (Oxford and Cambridge) took little interest in modern or scientific studies*."[283] The fact that many, if not most, of the 19th century advances discussed in this book nevertheless occurred in England can perhaps best be explained by Mumford's view that "*the detailed history of the steam engine, the railroad, the textile mill and the iron ship, could be written without more than a passing reference to the scientific work of that period*."[284] As Chapter 19b will demonstrate, internal-combustion engines could readily be added to Mumford's list. Mumford does see the dominant influence of science – his *neotechnic* phase –

beginning in the 1830s and in his Chapter 5(p238) he points to the importance of the many scientific discoveries of Faraday discussed in Chapter 15 below.

Some commentators have missed a key aspect of this debate. For example, Peters claims that the works of Euler on column buckling (Chapters 5d and 7a) and of Charles Coulomb[285] on earth pressure "*were all framed in mathematical language and therefore beyond the grasp of practitioners*."[286] However the mathematics of the key structural processes of bending and buckling were quite simple. The difficult step in each case was not the mathematical language but grasping the concept that the mathematics described. In buckling, it was that the singularity (or *eigenvalue*) in the resulting equation was not a computational nuisance, but an alternative definition of buckling. Similarly, bending theory becomes easy once the simple assumption that plane sections remain plane is explained.[287] Standard British bridge design texts in the late 1850s were still misrepresenting this basic aspect of beam behaviour.[288]

In the post-Revolutionary War period, American technical practice followed France rather than Britain. An American 19th century reviewer noted "*the progress of engineering science in America during 1775 to 1825 was largely directed by Frenchmen who came to this country at the solicitation of our Government.*"[289] Later, as the period of railway expansion demanded more and better bridges, Steinman observes that "*American youth flocked to Germany by the score to study at the font of scientific wisdom*".[290] Two of the greatest of the 19th century bridges were John Roebling's[291] 1855 Niagara Suspension Bridge and his Brooklyn Bridge in New York, completed in 1883 (Fig. 35). Although Roebling was a generation older than Eiffel (Chapter 7c and Fig. 34), the two men were producing their greatest works in the same period.

Roebling was born in Germany and studied engineering, architecture and mathematics at the Royal Polytechnic School (*Königliche Gewerbeakademie)* in Berlin before heading to America and the challenges of the New World in 1831. He built his first suspension bridge in 1845 and began work on Brooklyn Bridge in 1868. He died of tetanus contracted at the bridge site in 1869 and the bridge was completed by his son Washington Roebling. Washington had graduated in 1857 from Rensselaer Polytechnic Institute in Troy, New York State. Following the French and German models described above, it was founded in 1824 and was America's first engineering school. It is the oldest technical university in the English-speaking world.

Given the above situation, it is not surprising that none of the British heroes of the Industrial Revolution had a technical or scientific education, whereas most of the continental Europeans did. As the scene changed from revolution to implementation in the second half of the 19th century, this difference increasingly disadvantaged Britain. The case was well illustrated using the example of structural engineering and bridge design given in Chapter 7a.

Fig. 35 Roeblings' 1883 Brooklyn Bridge over the East River in New York. The 486 m main span was the longest in the world until 1902. Photo by user:postdlf at Wikipedia Commons.

British inventors and not British scientists dominated the first half of the Industrial Revolution (and, of course, engineering as profession did not exist in Britain during much of this period). It has been argued that the dominance of inventors was partly a consequence of Britain's early adoption of the patent system which, in an otherwise aristocratic society, gave the fruits of invention to the individual inventors and not to their wealthy patrons and employers.[292] It also gave substance to the view that, in their own right, practical ideas could be valuable property. Without patents, inventors could only profit by their personal use of their invention. The patent system allowed inventors to profit when anyone used their invention. The British system had begun in 1331 with '*letters patent*' and codified into a useful form early in the 18th century. On the other hand, most European patent systems did not reach a similar level until the 1840s. In his review of the sources of modern prosperity, Bernstein[293] highlights intellectual (as well as physical) property rights as one of the four prerequisites for economic growth.

It will be seen in Chapters 9 onwards that there was an absence of British inventors leading any of the post-steam engine developments. It is as if Britain was firmly focussed on the continuing development of the steam engine running on iron rails, without any incentive to explore alternatives. The manufacturing fruits of the Revolution – from the sewing machine to the car – were to be harvested elsewhere. Although Britain was a major exporter of steam locomotives in the second half of the century, the refinements in locomotive technology during that time were largely empirical and idiosyncratic and not based on a growing body of scientific and engineering knowledge. For example, the railway companies each had their own largely empirical preferences for the number of wheels, the size of the wheels, whether they were coupled together, and which were directly driven by the pistons (Chapter 11f). Relatively simple physical principles relating to the horizontal and vertical forces at the rail-wheel interfaces were frequently obscured by empirical personal preferences.

A related difference between the later development of steam and internal-combustion was that the steam locomotive advancements occurred within the British operating companies. They kept their initiatives closely guarded secrets and there were also a number of filters between actual locomotive performance in the rail network and the operational outcomes reported back to the locomotive builders. Personal rewards came from bureaucratic competence rather than from innovative skills. On the other hand, later Chapters will show that the new internal-combustion vehicles were produced by small entrepreneurial companies in Germany and France and in direct and measurable competition with each other. In abstract at least, the operators could observe and make rational decisions without considering company loyalty to a particular vehicle producer.

A somewhat separate issue was that, as discussed above, the overwhelming British contribution in the first half of the 19th century had been the result of contributions by extraordinarily gifted individuals. British education well into the 19th century continued to reflect a view that talent would rise to the surface of its own accord, almost in spite of the educational process. This approach missed the key point that, as the technology of the Industrial Revolution broadened and deepened in its interactions and impacts, there was an increasing need for teams of skilled engineers and scientists capable of understanding the processes that they were observing.

The educational systems of Germany, France and the United States were to prove far more capable of supplying such people than was the idiosyncratic British process. In 1855 William Fairbairn (Chapter 5c) was a juror at the Paris Exhibition. The role of these Exhibitions in promoting new technology – particularly internal-combustion engines – is discussed in Chapters 16 onwards. As a result of the experience he wrote: "*the French and Germans are in advance of us in theoretical knowledge of the principles of the higher branches of industrial art; and I think this arises from the greater facilities afforded by the institutions of those countries for instruction in chemical and mechanical science*."[294]

From the 1860s further shrewd British observers had been noting that neglect by Britain of technical education was a cause of its then current complacency and lack of major industrial momentum and leadership. The concern was formally raised in British Parliamentary Papers in 1868.[295] Nevertheless, the gap widened in the last quarter of the century when Germany changed its polytechnic schools into Technical Universities.[296] France, Germany and America maintained their lead over Britain in chemical and mechanical science for the remainder of the century.[297] In a subsequent monumental review of the history of technology, a group of British experts wrote "*A large share in the responsibility for the relative decline of Britain in the second half of the 19th century must certainly be ascribed to those who failed to see the importance of education, and above all of technical education.*"[298]

Certainly, in the 1850s Britain led the world in most aspects of technology and benefitted materially from this leadership. Nevertheless, as noted above, increasingly it devoted less of its educational efforts to technical education than did competitive economies such as Germany, France, Holland, Belgium, Scandinavia and America. There are no plausible explanations for this. At the best, one British commentator has described it as due to "*complex societal reasons*".[299]

Some of the earlier examples of education and training in the 19th century have been based on structural engineering as a relevant case study (and perhaps because it was the author's initial professional discipline). However to compare further the British apprenticeships with the system that had evolved in Europe during the 19th century, the emphasis must be changed to reflect the fact that this was the era in which steam-powered locomotives and cars powered by internal-combustion were developed. Both these engine types were based on taking energy created by combustion of a fuel and using it to produce power, with heat being a fundamental part of the process.

At the beginning of the 19th century the most common scientific explanation of heat was based on the view that heat was itself a substance – the *phlogiston* of the 17th century had been elevated to a new substance called *caloric* – and this viewpoint only slowly disappeared during the 19th century as the understanding of *thermodynamics* gradually developed.

The first doubts were sown by a nice coincidence as the greater availability of cast iron (Chapter 5c) led to the production of cast iron cannons with bored barrels. As Chapter 16 will demonstrate, this technology was soon used for producing the close-tolerance cylinders needed for the new piston-based engines. Benjamin Thompson[300] noted that the work of the boring machine raised the temperature of the cannon in a way that could not be explained by the existing caloric theory of heat. There were then many decades of scientific work directed at replacing the caloric theory, measuring heat and relating heat to the work used to create it. The major contributor to this work was James Joule[301]. It culminated in an agreed value for the mechanical equivalent of heat and was the basis for First Law of Thermodynamics, first postulated in 1850. Indeed in 1849, William Thompson[302] had proposed a new word "*thermodynamics*" in a paper of his on the "*efficiency*" of steam engines. In this context, efficiency was a very useful and deceptively simple term, calculated by dividing a machine's energy output by the energy used to produce that output.

The father of thermodynamics is retrospectively seen as Nicolas Léonard Sadi Carnot,[303] a French military engineer whose 1824 monograph defined the maximum efficiency of heat engines and was subsequently the basis of the second law of thermodynamics.[304] This law implies that no heat engine can be 100% efficient and was introduced at a time when steam engines had an efficiency of under 5%. At a wider level Carnot proposed the more esoteric concept of *entropy* which implies that all processes involve some permanent losses and therefore cannot be reversed. The word was coined by Rudolf Clausius[305] in 1865 and the theory introduced by him and by William Thompson, James Clerk Maxwell and others underlies our growing realisation over a century later of the fact that all the resources we consume are ultimately and inevitably exhaustible (see also Chapter 15b).

In his 1824 monograph Carnot perceptively remarked:[306]

> "*already the steam engine works our mines, impels our ships, excavates our ports and rivers, forges iron, fashions wood, grinds grain, spins and weaves our clothes, transports the heaviest burdens, etc. .. yet its theory is*

> *still very little understood, and the attempts to improve it are still directed almost by chance.*"

It remained "*little understood*" until the 1850s when people finally accepted that heat was a form of energy and not some mysterious substance that flowed by arcane rules. By this time there were a myriad of steam engines around the world producing useful power.

The operation of both steam and internal-combustion engines could be explained by laws of thermodynamics but it will be evident from latter Chapters onwards that both these engine classes were not a consequence of scientific and mathematical revelation, but instead the result of extremely clever and ingenious engineering – for steam in England in the first half of the century and for internal-combustion in Germany in the second half of the century. An exception is that Rudolph Diesel (Chapter 16b) who invented a major new form of the internal-combustion engine, was influenced by Carnot's definition of heat engine efficiency.

In the context of this Chapter, the supportive educational and vocational development of the poor but talented Gottlieb Daimler to be described in Chapter 14b stands out like a beacon and the circumstances which made it possible are one reason for the preponderance of major technical advances which occurred in the second half of the 19th century in a small region of south-western Germany, mapped below in Fig. 85. Daimler's major competitor, Karl Benz, had also received a similar intense education in both engineering theory and engineering practice (Chapter 14c). The region had been under French control in the first portion of the century and adopted the French education system of a high quality and comprehensive general education followed by detailed teaching and training in specific areas such as mechanical engineering. This training was usually associated with local industries which had been active in such areas as clock-making, coach building and the production of armaments.

But education and training are not the only requirements for the creation of successful innovators. The concentration of the development of the internal-combustion engine in south-western Germany was also bearing practical fruit and the locals were particularly alive to its commercial potentials. Inherently talented men such as Daimler and Benz were at the right place and at the right time. The words of the perceptive French commentator Raymond Bruckberger in describing Henry Ford could easily be also applied to Daimler and Benz.

> *"The strength of his genius lay in his passion for anything mechanical, for any kind of machine, from the movement of a watch to a locomotive. Ford said of his childhood years "My toys were all tools – they still are." And "Machines are to a mechanic what books are to a writer. He gets his ideas*

> *from them, and if he has any brains he will apply those ideas." He was a born mechanic."*[307]

What would people such as Ford and Daimler and Benz have become if they had been born in a less entrepreneurial location or a century earlier or a century later?

There will be a number of further opportunities throughout the text to contemplate why transport development switched so dramatically in mid-century from Britain to Europe. The British approach was certainly somewhat myopic, as epitomised by an otherwise excellent book about the "transport revolution" published in 1984 by British author P. J. Ransom[308]. It covers the period 1750-1850 and its final chapter on steam railways and ships is incredibly titled "*The transport revolution completed*".

To complete this review, it is appropriate to note that many have looked more broadly at the reasons why Britain lost its technical leadership in the second half of the 19th century and found further reasons. Externally, Britain did tend to increasingly devote its financial and physical resources to securing wealth overseas. Diamond mining in South Africa would be one example. Internally, there was a relative decline in work-force productivity, process innovation and managerial skills as many industries moved into a comfortable second-generation and lost the dynamics of competing for new markets with new products. The second generation also brought with it employees more conscious of disparities in the way the wealth they created was subsequently distributed.

PART 8

STEAM:
A NEW SOURCE OF USEFUL POWER

CHAPTER 8

STEAM:
A NEW SOURCE OF USEFUL POWER

The use of fire to provide useful heat was an early human innovation. Fires can readily produce enough heat to boil water and then convert the liquid into steam. Steam is a quite lively substance as, if not subjected to restraining pressure, it occupies some 1700 times the volume of the original quantity of water. Thus the pressure of the steam in a lidded container is often sufficient to blow the lid off the container. Consequently, jets of pressured steam from water boiling in a closed container had long tempted inventors seeking usable power. The first simple devices utilised the reactive forces due to steam jets escaping from small holes in the container. The main technical need was for containers that could withstand the increased temperatures and relatively high pressures. Over the ages, such jets were used to produce motion in toys carrying the fire, the boiling water and an aligned jet for the escaping steam. The next development blew the jets of steam on to vanes attached to an axle of the vehicle carrying the boiler. The processes were not efficient as much of the usable steam pressure was lost as the jet expanded in the open air.

A more sophisticated 17th century application used the steam jet to turn the blades of an enclosed turbine linked by cog wheels to an axle. The efficiencies of the process were still too low for the device to be effective, but a way forward was suggested in 1827 when Benoit Fourneyron[309] in France operated a turbine receiving water from a penstock. Fourneyron's turbine operated at 85% efficiency compared with the 20% efficiency of a water-wheel.[310] Steam is more difficult to manage than water and steam-powered reciprocating engines have very low efficiencies of about 10%. This led to continuing efforts to improve the detailed design and construction of steam turbines culminating in a design arrangement produced by Charles Parsons[311] in 1884 which had an efficiency of about 30%. Today most thermal power stations use steam turbines to convert heat to electric power.

The pressure produced by the formation of steam can also be used to cause pressure on the face of a piston which forms one end of a closed cylinder. Initially this method was not feasible as it was not possible to produce containment vessels which would permit the development of pressures high enough for the piston movement to be of practical use. However, if the steam in a sealed containment vessel is cooled, the pressure drops below atmospheric pressure in the container as the steam condenses (shrinks) to water. This sub-atmospheric pressure is called a *partial vacuum* and can also be thought of as *negative pressure* or *suction*. Suction pressures are relatively low and so do not demand tight-fitting containers. The most useful initial application of the suction process was to seal cylinders.

Using suction to do useful work by moving a piston was first suggested by Leonardo da Vinci in about 1500 at the beginning of the Scientific Revolution.[312] The da Vinci device did not use steam. Instead, fuel was burnt in a cylinder closed at one end by a piston. This produced expansive combustion gases. The cylinder was then rapidly cooled and, as the combustion gases condensed to liquid, they caused a suction which drew the piston into the cylinder. The device is now known as the "*fire engine*" and, as a concept, was the first internal-combustion engine. Engines using this suction process to utilise the atmospheric pressure of about 100 kPa were called *atmospheric engines*. The pushing force on the piston is the atmospheric pressure on its outer face, less the pressure of the partial vacuum on the inner face.

A simpler hypothetical form of the da Vinci engine was the gunpowder engine which fired gunpowder within a vertical cylinder closed by a piston. The explosion forced the piston in the vertical "cannon" upwards, with work being done as the piston went upwards and then returned downwards due to gravity. Pressures due to explosions can be relatively high and thus difficult to manage.

Although the Dutch scientist Christiaan Huygens[313] was unaware of the da Vinci *fire engine*, in 1673 he used an extension of the method in which the vacuum was produced by exploding gunpowder in one part of the cylinder, forcing a piston to push air from the cylinder. The air left via a one-way valve, creating a partial vacuum in the cylinder, which then sucked the piston back, producing a cycle of useful work. Some questionable accounts state that it was used to pump water from the Seine at Marly to Versailles Palace, about 5 km to the south. The device was not a practical success and stayed as a scientific curiosity for another 130 years. Some have suggested that the device did not rely on the vacuum and was more like a vertical cannon, however Hardenberg strongly disputes this view.[314]

Denis Papin[315] was a protégé of Huygens, but he came to favour the use steam rather than gunpowder.[316] In 1690 improved manufacturing technology allowed Papin to develop a subtle and useful device which produced the suction pressure by cooling the steam back to water. The operation involved surrounding the water-filled cylinder with a furnace to turn the water into steam, most of which was allowed to escape from the cylinder. The steam valve was closed and then the furnace was removed. As the cylinder slowly cooled, the remaining steam condensed back to water and a partial vacuum was created in the cylinder, sucking the piston back and thus doing usable work. The process was somewhat ponderous and took 3 minutes to produce one piston cycle. The advantage of the device was that the pressure produced by suction was a number of times greater than expansive pressures that could be developed by steam using the containment systems then available. Papin also tried using steam containment pressure to help push the piston outwards.

A decade later in 1698 Thomas Savery[317] in Devon took the suction idea further and produced a practical water pump for shallow mines. His device placed the steam directly on the water to be raised, thus avoiding the need for a piston. It is simplest to explain the process using a slightly later version of the engine shown in Fig. 36. The cylinder (B) is separated from the furnace (A) which boils water and produces steam which is taken to the cylinder by a tube and valve (V). This forces the piston (P) upwards and the bucket (attached to F) downwards. Aided by cold water (C) through another tube and valve (V'), the steam then cools and condenses. The resulting partial vacuum sucks the piston downwards. The piston pulls D down and the beam raises F and its attached bucket full of water or ore. The process produced about 700 W and was widely used for pumping water in small drainage operations.

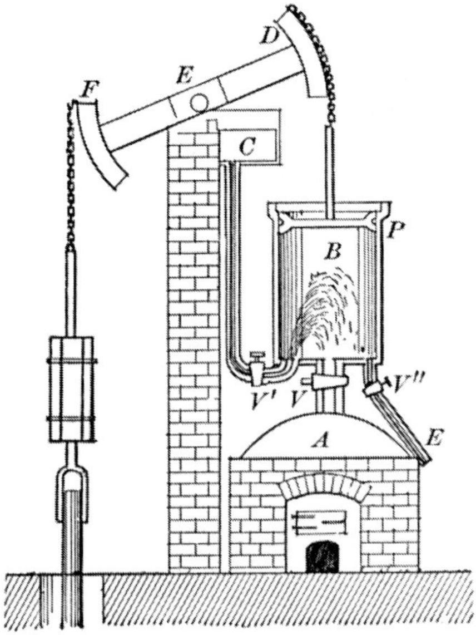

Fig. 36 Principles of the atmospheric beam engine (drawing said to be based on a Watt engine). Wikipedia Commons, from Black and Davis.[318]

In 1712 Thomas Newcomen[319] at Dudley near Birmingham dramatically improved the efficiency of this process with his cylinders, pistons and boilers all made from cast iron (Chapter 6). The boilers converting water to steam were based on efficient heat-transfer arrangements developed by the brewing industry. The boiler was separated from the cylinder and this allowed more effective use of cold water to condense the steam and produce suction.

The Newcomen engine was the world's first successful and practical steam engine and the underlying system came to be widely used. For most of the 18th century, the pistons transmitted power by moving large pivoted beams (DEF in Fig. 36) which produced linear motion in rods attached to them. This was their most visible feature and they were commonly called *beam engines*. They initially produced about 3.5 kW which was sufficient for many mine drainage applications. In 1767, fifty-seven Newcomen engines were working in the coalfields in north-eastern England.[320] Their power output was then about 10 kW. In dewatering mines, it was estimated that one Newcomen engine replaced about 25 horses. Over a thousand Newcomen engines were built.

The beam engine was relatively large and cumbersome. The alternative *crank* system for converting linear movement of a piston into rotary motion by attaching it to eccentric cranks in a rotating shaft had been used since at least Roman times, but for relatively low-speed applications. James Pickard,[321] a Birmingham mill operator, patented the use of a cranked drive to take power from a steam engine in 1780. Crankshafts and bearings suitable for the higher revolving speeds of steam engines began appearing in the 1780s, as machining technology improved. In a short period, this enabled steam power to be used for a wide range of new industrial and agricultural applications.

By today's standards, these engines were very large, had a low power to weight ratio, and a mechanical efficiency of well under 5%. A major reason for the inefficiency was the need to alternatively heat and cool the cylinder on each piston stroke. This inefficiency did not preclude their use at mine sites, but their size inhibited their use for transport applications.

Nicholas Cugnot[322] had served as a French military engineer in Austria. He had seen Newcomen steam engines and in Paris in the late 1760s he built a 5 t steam-powered tractor intended to haul cannons around the city's defensive walls (Fig. 37). It was the first realistic steam-powered vehicle. To provide power, Cugnot had reverted to a Huygens arrangement using only the expansive pressure of the steam produced by a separate copper boiler (on the right in Fig. 37). The steam was taken by a tube to a pair of cylinders, each with a 330 mm bore, a 300 mm piston stroke and a 26 L capacity. Two cylinders were used in order to provide the vehicle with continuous movement. Nevertheless, the engine could only operate continuously for about 15 minutes.[323]

Fig. 37 First and only public outing for Cugnot's steam-powered vehicle (1771). Two vertical arms can be seen: the one on the right is the ratcheted piston; the one on the left operates the steam valve. The cannon was suspended from the two heavy wooden beams on the left. via Wikipedia Commons

Instead of a beam or a crankshaft, Cugnot produced rotary motion by using two cylinder pistons working through two ratchet mechanisms on either side of the single wheel to produce a travel speed of about 10 km/h. This single wheel was also used for turning the vehicle and so, not surprisingly, steering was very difficult. There were a few short test runs in 1769-70 and in 1771, on its first and only public outing, the tractor demolished the walls it was intended to protect. The vehicle was later rebuilt and placed in the Musée des Arts et Métiers, rue Saint-Martin, Paris, where it remains today.

James Watt[324] had trained as an instrument maker in Glasgow and was working in that role at Glasgow University when in 1763 his employers asked him to repair a Newcomen engine. He repaired the engine but was unimpressed by its performance. He realised that a greater power output was possible if the repeated heating and cooling of the cylinder could be avoided. In 1769 he invented and patented a separate condenser which allowed the condenser to be kept cold and the cylinder to be kept hot. This at least doubled the efficiency of the Newcomen process and the resulting reduction in engine size made steam power a more practical proposition. Nevertheless, the overall efficiency was less than 4% which mean that the engines still had to be quite large to produce useful amounts of power.

In his work Watt was far more an inventor than a theoretical scientist. However, during his time at Glasgow he had been aware of the work of a Glasgow professor, Joseph Black, who had just unveiled the concept of *latent heat* which provides a theoretical basis for the efficacy of Watt's condensers. In 1775 Watt formed a partnership with Matthew Boulton[325] to exploit his invention. Boulton was an active, successful and well-known businessman, "*steeped in mechanical lore and enthusiastic about mechanical invention*".[326] He had made his fortune mass-producing cheap jewellery and his family subsequently became part of England's landed gentry, owning an estate of over 3000 ha of land in the Cotswolds.[327] Boulton's Birmingham metal-working facilities were able to bore cast iron cylinders to the accuracy required by Watt's new designs. Up till then Watt had used cylinders made from tin to obtain closer fits between piston and cylinder (Chapter 5c). In 1776 Boulton told a visitor to his facility, the biographer James Boswell, "*I sell here, sir, what all the world desires to have – power.*"[328]

Watt's inventions continued and in 1782 he produced and patented an engine which used the expansive pressures of heated steam on one side of the cylinder, and the atmospheric engine's suction pressures from cooling and condensing steam on the other side. This so-called *double-acting* arrangement was first used commercially in 1785 and proved far more efficient than its predecessors, providing a power output of about 8 kW.

Watt was indisputably at the pinnacle of steam engine developers. He also invented parallel motion linkages to capture power from both directions of the piston in double-acting engines, planetary gearing to replace both the crank and beam engines, and centrifugal speed governors to control engine revolutions. Boulton also remained active and in 1795 he built the Soho Foundry in nearby Smethwick which was probably the world's first factory. Indeed, Birmingham was the centre of factory production during the Industrial Revolution and was then Britain's third largest city.

Steam engine development continued, particularly after the key *Boulton and Watt* patent expired in 1800. By then the firm had made over 500 engines, with the majority producing a rotary power output, typically of about 18 kW. These turn-of-the-century engines were relatively heavy. One was built in 1812 to pump water for the Kennet and Avon Canal near Great Bedwyn in Wiltshire. The single-acting condensing beam engine has been in service ever since, although it now only operates at weekends. *Boulton and Watt* continued producing fine engines until firm closed in 1895.

Watt disliked the idea of high-pressure engines. However, in 1801 in Philadelphia Oliver Evans[329], who had trained as a wheelwright, began the development of high-pressure steam engines which had power-to-mass ratios noticeably superior to the Boulton and Watt engines. In 1804 Evans built a steam-driven flat boat (cum wagon) – the *Orukter Amphibolis* – which weighed over 20 t. Details of its use are sketchy – it may have operated briefly as a steam-powered dredge in Philadelphia in 1805 and have made two trips as a road wagon, before its wheels collapsed under the vehicle's self-weight.

Watt – whose firm had the bulk of the market – actively discouraged the use of his engines to power vehicles.[330] Apart from the steering issue, two other core technical problems with using heavy engines on vehicles were associated with the gravitational traction forces produced on even modest grades and the effect of the wheels on the travelled way. As mentioned in the earlier discussion of animal-drawn vehicles in Chapters 1 & 2, without pneumatic tyres, wheels had to be made very large to minimise effective contact pressures between the tyre and the road (Fig. 2). To prevent very rapid wear, the wheel rims were faced with iron plates or with the heads of iron pins hammered radially into the rims. This meant that even on well-made road surfaces, the slightest raised local prominence would be subjected to very high local contact pressures. This caused rapid wear and distortion of the road surfaces. Even modern roads are quickly damaged by wheels with "rigid" rims. At a slightly larger scale, the wheel loads cause a pavement to "dish" under load such that extra power is consumed by the wheels continually climbing out of their contact dish.[331]

These traction problems meant that the initial practical transport applications of steam power occurred in the first decade of the 19th century when it was used on vessels hauling barges along rivers and canals (see Chapter 9).

The potential of steam power was certainly being realised by a prophetic few and it is appropriate to leave the 18th century with this verse written by Erasmus Darwin, Charles Darwin's grandfather, in 1791 in his "Botanic garden":

Soon shall thy arm, Unconquered Steam! A far
Drag the slow barge, or drive the rapid car;
Or on wide-waving wings expanded bear
The flying chariot through the fields of air,
- Fair crews triumphant, leaning from above,
Shall wave their fluttering 'kerchiefs as they move;
Or warrior bands alarm the gaping crowd,
And armies shrink beneath the shadowy cloud.

The idea was not a wild poetic fancy as Erasmus Darwin was a close personal friend of Mathew Boulton and they both belonged to one of the earliest scientific-cum-engineering bodies, the Society for the Encouragement of Arts, Manufacturing and Commerce. Indeed, although Darwin was a medical doctor he was also a productive amateur mechanic who had designed pumps and steering and steam-power systems for carriages.[332]

If there was a single event that could be said to have caused the Industrial Revolution it would be the development of useful steam power. As the next chapter shows, steam power changed from being a curiosity associated with clever toys and social conveniences to being the working, pulsing core of the Industrial Revolution. This was due, of course, to ingenious inventions but these inventions were no longer primarily demonstrations of personal cleverness but rather they were commercial responses to rising demands for greater production and consumption and wealth accumulation.

Initially, the commercial demand for power had come mainly from mine-owners who needed it to pump water from their ever-deepening mines. The core problem was how to provide the power. There had been some success with using water to turn wheels, but this was limited to specific localities and seasonal times. When steam became increasingly feasible as a major power source there was a crescendo of demands as the commercial usefulness of the new devices came to be realised. Change was no longer slow and steady. Quantum jumps were occurring. Long-standing agrarian life styles and inefficient production processes were often brutally pushed aside by market demands. Some of these detrimental impacts were already mentioned Chapter 6.

The development of iron and the consequential development of more powerful engines were partners in the overture to the Revolution. The interactive nature of the processes at work is made evident by the fact that none of the iron industry developments described in Chapter 5 could have occurred at the pace and magnitude described without the new availability of steam power. To produce useful amounts of energetic steam required large quantities of fuel. To make the engines and to transport the fuel required more iron, and iron-making required more steam *et seq*. It was an incestuous and self-fulfilling process. Thus coal fuelled the Industrial Revolution by not only powering the new machines, but by also providing an economic incentive to develop more of these powerful new devices.

The only viable way to produce steam was by burning coal (or charcoal). There were many coalfields throughout Europe and North America and the coal varied in quality and accessibility. As discussed in Chapter 6, England was particularly blessed with large deposits of good quality coal close to points where the coal could be usefully used. As Carnot observed from France in 1824 "*coal-mining increased ten-fold in England since the invention of the steam engine*".[333]

A major German commentator has noted that in England the relatively centralised coal industry made coal an everyday reality, whereas in France the coal industry was "*scattered throughout the country . and the French were unable to perceive coal as the endlessly available fuel*".[334] He illustrated this with two opposing opinions. In England, Adam Smith and others had calculated that a horse ate as much as 8 men. As there were a million coach horses, moving to steam power would release food for 8 million men. The French on the hand observed that coal deposits were exhaustible, but horses and their food could be continually renewed.

The following table shows coal production in the two countries during the first half of the 19th century and dramatically illustrates the difference:

	Annual coal production, Mt			
year	1816	1836	1846	1851
England	16	30	44	57
France	1	2	5	7

Many commentators from Kuznets[335] onwards have observed that rapidly developing industries do not grow rapidly indefinitely, but are overtaken by new, consequential industries that enter their own stage of rapid development. The following Chapters on machine-powered transport will clearly demonstrate this process and it has already been encountered in the 19th century development of iron and steel making and using discussed in Chapter 5 and on the growing use of steam power described in this Chapter.

PART 9

THE TRANSPORT REVOLUTION IS UNDERWAY

CHAPTER 9A

THE DEMAND FOR TRANSPORT

So, as the 19th century began, rapidly increasing numbers of people and goods were moving about with purpose and profit. Some communities had changed from the subsistence economies of medieval times to centres of specialist production, sophisticated trading and conspicuous consumption. Others looked on with envy.

The transport of items between producers and consumers is an essential part of the process of production and consumption. However, as earlier Chapters demonstrated, there was only modest incremental change in transport capability during the 18th century. Travel speeds for cargo and for most people had not improved at all, and some speeds had declined as the amount of traffic increased. Freight capacity had only improved in the limited circumstances where the use of ships and barges was viable. This is evident in the significant 18th century investment in canals, but the canal barges were still horse-drawn and slowly followed circuitous routes. Large loads could be moved, but at walking speed or less. Nevertheless, by 1815 Britain had about 5 000 km of canals, mainly built using cheap Irish labour. The growth of British coastal shipping as the 19th century opened was discussed in Chapter 1c

The demand for major new transport services was certainly there and the increasing availability of iron and steam-power provided the catalysts which enabled that demand to be met and in themselves produced new demands. In addition, the mining industry had shown conservative communities just how useful the new steam-powered technologies could be. Most early mines had been quarries or horizontal shafts driven into hillsides. With the advent of steam-powered pumps and hoists it was possible to use vertical mine shafts to seek valuable ores well below the ground surface.

The two factors fed on each other, for mines were the forerunner of many new industries which would, like mines, produce their large outputs at a single point. Moreover, their product was often consumed at another single point. It was therefore relatively easy to justify major expenditure on freight transport between these two points. On the other hand, the existing farm products were collected from disparate producers, processed in a range of places, and then distributed to disparate customers. There had been far less organised incentive to invest in transport facilities for the farm-to-market system although, as discussed in Chapter 3a, the need had been the basis for Arthur Young's continuing efforts to improve British roads.

CHAPTER 9B

ALMOST-USABLE STEAM-POWERED VEHICLES

Chapter 8 described how the Watt engine had provided a major advance in the use of steam to produce power. There had been a few trials of the engine to power a vehicle. In particular, in Cornwall in 1784 Watt's assistant, William Murdock,[336] mounted a small engine on a miniature chassis. The single vertical piston delivered power to the driving wheels via a beam and crank arrangement. The device was only about 300 mm high (Fig. 38) and first operated on the floor of Murdock's house in Redruth. Murdock never built a full-sized vehicle, perhaps discouraged by Watt who subsequently patented his own vehicle design. Watt nevertheless preferred to devote his attention to developing his stationary engines, which were in increasingly high demand at mine sites.

Fig. 38 William Murdock's first practical steam-powered vehicle. The device was about 300 mm high. With permission of the Science Museum London (10221419).

Richard Trevithick[337] was a Cornish blacksmith turned mining engineer who had had little formal education. He was familiar with Watt's work and had been directly influenced by Murdock. He used the improved metals and manufacturing techniques available at the beginning of the 19th century to produce high-pressure steam engines which avoided the Watt's patent and had better efficiencies (at about 15%) and higher power-to-mass ratios than their predecessors. This was partly because his better-fitting cylinders and pistons allowed him to drive the pistons using the pressure of confined steam at about 300 kPa, which was far higher than the 100 kPa suction pressure caused by condensation. Although this demanded much stronger pressure vessels, there was no need for condensers and the boilers and pistons could be placed in a more compact arrangement resulting in much smaller engines for the same power output. The puff of steam released at the end of each piston stroke led to the new engines being known as "*puffers*". Trevithick also introduced the use of the exhaust steam to pre-heat the boiler water and provide a draught for the furnace.

The improved power-to-mass ratio of his new engines raised Trevithick's interest in steam-powered vehicles. He built his first such vehicle in Camborne in Cornwall in 1801. The piston was directly coupled to the driving wheels of the vehicle. In early December of that year, with twelve passengers on board, "*Captain Dick*" – as he was then known – drove his vehicle at about 20 km/h along a flat road and then travelled up a 5% grade. The paradigm shift that this event caused is curiously illustrated by the fact that a week later Trevithick and his helpers parked their new coal-burning invention in a barn whilst they had a celebratory Christmas Eve meal of roast goose at a local inn. Both barn and carriage were destroyed by the ensuing fire.[338] The event did not lessen Trevithick's enthusiasm and drive and he was the first to use horizontal cylinders.

In 1803 Trevithick manufactured another steam-powered carriage in Camborne which he used to transport some passengers on London streets at speeds of about 15 km/h. It was the first practical self-powered (or *motorised*) vehicle (Fig. 39). However, it was uncomfortable and it was soon obvious that it could not compete commercially with the horse-drawn carriages then in common use (Figs 11 & 12). In 1804 the carriage lost traction between its wheels and the road surface when it was negotiating a steep decline.[339] Trevithick's response to this was to abandon road-based steam carriages and the outcome of that decision is discussed in the next Chapter.

Fig. 39 Trevithick's 1803 steam-powered carriage. It had a single horizontal cylinder with the piston movement changed to rotary motion by the use of a crank. A flywheel ensured continuous motion. The drawing is the only existing representation. It is very idealised and shows no provision for carrying fuel and water or for loading passengers. The operating pressure was about 200 kPa. via Wikipedia Commons.

Despite such road transport dead-ends, stationary steam engine technology continued to develop at exceptional pace. In 1840 there were about 5 000 steam engines at work in Britain, 200 in France and almost 100 in Prussia.

CHAPTER 9C

TRAVELLING ON DEDICATED WAYS

As Trevithick found in 1804, there were seemingly insuperable problems with the common road systems described in Chapter 3. Indeed, soon after the invention of the wheeled ox-drawn wagon, the new wheels began cutting grooves – or ruts – into soft road surfaces. Grooved roads have been found in the remains of Roman Pompeii and in ancient roads in Mesopotamia, Malta and the Swiss Alps. The famous Oedipus story from ancient Greece involves chariots caught in wheel ruts on the Cleft Way between Thebes and Delphi.[340] As a consequence of such ubiquitous grooves, over millennia a wheel spacing of about 1.4 m has remained a constant for road and rail vehicles. The dimension was mainly a consequence of practical limits on the size of the wooden wagons and the harnessing of the hauling oxen.

Based perhaps upon these accidental events, the rut problem was turned into an advantage, with some early dedicated hauling systems operating with the wagon wheels running in manufactured ruts. The process not only guided the wagons but also ensured that their wheels ran on firm ground. It was a labour-intensive solution but required minimal technology.

A more technologically-intensive alternative was to operate wagons with flanged wheels running on raised rails.[341] Rectangular lengths of wood known as edge rails had been used since medieval times for hauling wagons in underground mines,[342] at collieries[343] and to link collieries and wharves, as in 18th century Newcastle.[344] These "*wagonways*" were subsequently extended to work-sites and wharves. Without such systems, the in situ surface would quickly become impassable due to the pressure from the hard rims of the wheels, the pounding hooves and the defecation of the animals hauling the wagons.

Ideally, the wagons also had wheels with flanges that maintained the lateral position of the wagon by using the vertical face of the edge rail. Initially the wheels had wooden flanges as iron-flanged wheels required the next generation of iron-making. It then became possible for wooden wheels to be made with an iron rim that included a flange. The next step was taken when Coalbrookdale ironworks produced a few small cast-iron wheels with integral flanges and larger iron-flanged wheels began appearing early in the 19th century (Fig. 40). Flanged wheels reduced the need for raised rails. Operating problems due to raised rails are discussed further in Chapter 12c.

Fig. 40 Coal wagon in northern England, c1815. Note flanged wheels on raised wooden rails. By 1880, there was about 300 km of rail track carrying horse-drawn wagons servicing the local needs of British collieries and ironworks. From [345].

The rails wore quickly and, as iron-making technology developed (Chapter 5c) this situation was improved, initially at Coalbrookdale in 1767, by using cast iron strips as a running-cum-wearing surface on the horizontal face of the wooden edge rails and then by using a complete cast iron edge rail. This latter approach was introduced at Dowlais in Wales in 1791 and at Middleton in Durham in 1797. However the rails were too narrow and wore grooves in the wagon wheels. Iron rails with a variety of cross-sections were tried England and Wales in the 1790s.[346] L-shaped sections were also used helped guide conventional wagon wheels.

In order to span larger distances between underlying supports, the rails had to be given more structural bending strength than could be achieved by a simple rectangular or L shaped cross-section. In the 1790s it was realised that the rails were more able to transfer heavy wheel loads to the ground below the rails by casting the rails with (1) a structurally efficient I cross-section with a high second moment of area and (2) a longitudinal fish-belly shape giving maximum strength between supports (Fig. 41). An early railway using iron fish-belly rails and cast-iron flanged wheels was operated in Leicestershire under horse power between Belvoir and the Grantham canal from 1815 to 1915.[347]

Fig. 41 Fish-belly rails, on stone blocks which were the common supporting system before transverse sleepers. by Chevin in Wikipedia Commons.

As discussed in Chapter 5, making lengths of iron by rolling rather than wroughting had been in limited use since the 16th century and in 1783 Henry Cort in Fareham in England took advantage of the new steam-based power sources and more malleable irons to begin the commercial production of iron bars by rolling. This avoided problems associated with the brittleness of the cast iron. Malleable wrought iron rails were introduced in 1805 and their ductility resulted in a far better service life than the more brittle cast iron rails. As rolling equipment developed further and malleable irons became more commonplace, rolling replaced wroughting as the best method for making long and ductile iron rails. Thus it became possible to produce more sophisticated rail shapes and better flanged wheels. The new rails and wheels could adequately both carry and guide loaded wagon wheels.

Nevertheless, in 1810 iron rails with a 40 mm square cross-section were being produced and had found favour with railway innovator George Stephenson (Chapter 9d) who used flanged iron wheels running on square iron rails made by John Birkinshaw[348] in Badlington, Northumberland. However, without closely spaced supports, the system could not carry the vertical wheel loads produced by the later and larger steam-powered locomotives.

The next advance – again driven by the mining industry – occurred in 1820 when Birkinshaw produced wrought iron rails up to 6 m in length, which were much easier to install and maintain than shorter rails. In 1821-5 George Stephenson used these rails for the Stockton to Darlington railway to be described in Chapter 9d. There are reports of similar products being produced in Germany in the same period.

As shown in Fig. 41, stone blocks (typically about 600 x 600 x 300 mm) at a spacing of about a metre were often used to support the rails and distribute their load to the underlying formation. Their position had to be stable enough to keep the rails to the required lateral spacing – or *gauge*.[349] (More specifically, *gauge* is the distance between the inside vertical edges of the pair of parallel rails). Stephenson used the block system to support the rails of his Stockton to Darlington line, compacting the underlying formation by dropping the blocks from a height that would produce a force equivalent to that expected from a passing locomotive.[350]

Not surprisingly, supporting rails on blocks did not prove satisfactory – for example, the blocks would crack or tilt under the inevitable off-centre loads. It then became common to support the rails on longitudinal timbers, called *sleepers* or ties or transoms which were old terms used for horizontal supports in buildings and ships. It was also necessary to use transverse sleepers at a spacing of about 5 m to maintain the rail gauge. Employing longitudinal sleepers meant that fish-bellied rails could no longer be used.

As rails improved structurally and could span greater distances between the supports it became obvious that a more effective overall solution was to use the sleepers transversely to both transmit load to the underlying formation and to maintain the rail gauge. The still-current reliance on transverse sleepers came into general usage during the 1840s, with a sleeper spacing of under 2 m. Nevertheless, placing rails on transverse sleepers is an inherently poor structural solution as it concentrates the wheel loads at specific points. This was one reason why the longitudinal method was continued by Isambard Brunel on the large gauge (2.1 m gauge) Great Western Railway.[351] He achieved transverse gauge stability by placing the longitudinal bearers on transverse sleepers at a spacing of about 5 m.

Flat-bottomed rails were introduced by Robert Stevens[352] in America in 1830 and made the attachment of rails to sleepers both simpler and more effective. In the late 1840s rails were joined longitudinally by fish-plate splices, thus greatly increasing ride quality.[353]. Rails with the current bulb heads came into use in the 1860s as steel-making and powerful rolling mills became available (Chapter 5d). Despite the growing availability of iron rails, local cost structures meant that the use of wooden rails was common throughout the world until well into the 1870s.[354]

The common railway practice of placing the sleepers on a layer of stone (ballast) placed over the formation broadly followed McAdam's then-current road-making practice (Chapter 4a) was (and still is) a far from optimal solution. First, it discouraged the production of a stiff and strong formation as any settlement of the track was corrected by later "*lifting*" the track and sleepers by placing more ballast beneath the sleepers. Second, the layer of stone was based on early macadam practice (Chapter 4a) and used relatively large pieces of broken (sharp-edged) stone. This had two problems. The layer was very porous so any rain-water passed straight through onto the natural formation which in most cases would be far weaker when wet rather than dry. The other problem was (and still is) that the inter-particle stresses between the sharp edges of the stones which form the ballast were very high and often led to rapid break-down of the stones.

Railways have consistently ignored the road-making lesson that it is better to use stone mixes containing a range of sizes so that voids in the layer of large stones are filled by smaller pieces of stone. This makes the layer far less porous and reduces inter-particle stresses.[355] Road-makers also appear to place more emphasis than their railway colleagues on using roadside drains to keep water out of the formation and to remove water from it.[356] It does seem to this admittedly biased observer that the rail industry placed all of its innovative efforts into the locomotives and the service they provided, and stayed with a relatively simple, unchanging and sub-optimal track system.

The first grooved rails were developed and rolled by Alphonse Loubat[357] in France 1852, following his experiences introducing street railways into New York in the 1830s. As will be discussed in Chapter 12c, street railways – also called streetcars, trolleys, trolley cars or trams – became commonplace in many large US cities. The grooved rails were of assistance to vehicles on demanding urban curves and where public pressure had forced operators of street railways to lower the rails so that they did not protrude above the street surface.

Many of the industrial applications for steam-powered heavy haulage were far from flat. Even in flat applications there was initial (misplaced) scepticism about the horizontal frictional forces that could be developed between iron wheels and iron rails. This was not finally resolved until Stephenson's time (Chapter 10b). In 1811 the scepticism had led to John Blenkinsop's[358] development of the rack-and-pinion railway in which a toothed powered wheel (the pinion) engaged with cogs on a trackside rail (the rack). The concerns about low friction were found to be excessive and the method was subsequently only used on extreme grades. Blenkinsop used a steam locomotive with two vertical cylinders working as beam engines (see Fig. 38) – the *Salamanca* – designed and built by Matthew Murray[359] in 1812. The work was for the Wylam Colliery near Newcastle.

An alternative approach was to share the task of producing horizontal traction forces by coupling other running wheels to the powered wheels. This technique was conceived by William Hedly[360] when working for Trevithick. He then introduced it in 1813 at the Wylam Colliery. It was the first railway to rely solely on wheel-rail contact for traction.[361] The adhesion was fine, but the locomotive was too heavy for the current rails and it only operated for short time. It was brought back into service in 1830 when stronger rails were available. Named *Puffing Billy*, the original engine is preserved in the Science Museum in London and is the oldest existing steam locomotive.

The vertical wheel loads on such locomotives were still too high and this soon led to use of the *bogie* system in which sets of wheels are carried on an independent sub-structure beneath the locomotive. The use of a leading four-wheeled bogie was introduced in America in 1832 to stabilise locomotives operating on uneven tracks.

There was also a major above-ground alternative to the use of ruts or rails. In 1758 Charles Brandling[362] built a special road at Middleton south of Leeds to carry horse-drawn wagons of coal into Leeds. In his wagonway the two wheel-path strips were paved with wood. An Act of Parliament was required to obtain right-of-way and it has been argued by some that this was the first formal railway, although the rail was a flat plate rather than a rail and subsequent similar systems were often called *plateway*s.

The plateways were found to greatly improve haulage operations at work sites. The advantages were twofold: they provided an all-weather path and the haulage capacity of a horse was significantly increased, often by a factor of 10. This was important as the Napoleonic Wars had caused a major increase in the price of horse fodder and thus of the costs of horse-drawn haulage. Another form of plateway was applied on long steep hills on public roads. Two rows of granite blocks were placed in the pavement to assist horse-drawn vehicles in their ascent of the steep gradient.[363]

As iron became more plentiful, the plateway technique was modified to take advantage of iron's superior properties. This first application occurred at Coalbrookdale in 1767 where the plateways typically used metre or so lengths of cast iron about 100 mm wide and 30mm thick. They were effectively pieces of pig iron (Chapter 5a). The associated iron-maker Richard Reynolds[364] then introduced L- shaped pieces which offered many of the advantages of a rail system without the need for a flanged wheel. There were growing demands from the mining and iron-making industries and these plateways were the preferred haulage approach for about 20 years. A further innovation occurred in Sheffield in 1780 when colliery superintendent John Curr[365] supported the flanged plates on transverse wooden beams thus predating the sleeper system for rail systems described above.[366]

By 1811 the Merthyr works in Wales had some 50 km of plateway in operation. Plateways were first extended beyond work-sites with the Surrey Iron Railway which operated between 1803 and 1846 as a horse-drawn freight service between Wandsworth and Croydon in Surrey. It used L shaped iron plates supported on stone blocks. Users provided their own wagons and horses and paid a toll to use the plateway. One use was to bring good road-making stone to London.[367] People were soon proposing a national network of plateways, very much akin to the subsequent railway network.[368]

A major advantage of the plateway was that a vehicle could leave the plateway and use the existing road system, thus integrating road and rail far more than did the subsequent railway. The disadvantage of the plateway was that the plates had very little structural strength and had to be supported at close intervals over most ground conditions. Support conditions would be increasingly important as the forthcoming steam locomotives were far heavier than any preceding transport vehicles.

In the 1830s the New York and Harlem River Railroad built some lengths of street track using stone slabs in which parallel rut-like grooves had been cut. This allowed their horse-drawn carriages to run smoothly and in an orderly way. The tracks could also be used by private carriages with the appropriate wheel and rim widths.[369]

The initial development of rail systems described above is yet another example of the major impact that the demands of and potential rewards from mining have had on the development of technology generally and of transport in particular. This repeats the incestuous theme raised earlier, for many of the products of mining were used to build and operate the new transport systems.

CHAPTER 9D

THE RAILWAY REVOLUTION

Trevithick's problems with road-based steam carriages were described in Chapter 9b. As a consequence, he concentrated on building a number of steam-powered haulage vehicles intended for hauling wagons at iron-works and collieries. The first was built for the Pen-y-Darren iron-works at Merthyr in Wales. It was approved by an Act of Parliament in 1803 and in its first outing in February 1804 it hauled coal over a 16 km track from Pen-y-Darren to a wharf on the Glamorganshire canal.[370] It was thus the world's first rail-based steam-powered vehicle. Some have claimed that it was the world's first railway.[371] However, the discussion in Chapter 9c noted that the Middleton *plateway* operated from 1758 and the Surrey Iron Railway was approved in an 1801 Act.

The track for the new Pen-y-Darren system had been built by the Coalbrookdale Company and this first self-powered train is often called the *Coalbrookdale Locomotive* (Fig.42). Before describing the vehicle it is useful to consider its name. As noted in Chapter 5b, Coalbrookdale is seen by many as the birthplace of the Industrial Revolution. The *Coalbrookdale Company* was founded in 1709 and is still in operation today. The term *locomotive engine* was coined at the time to distinguish the new engines from the stationary steam engines by then becoming common at mines and elsewhere. "*Loco*" is from the Latin for "place" and "*motive*" means "move". The word *locomotion* to describe moving from one place to another had been in use in the 17th century and so a *locomotive engine* was seen at the time as an engine for moving items from place to place.

The Coalbrookdale Locomotive had flat wheels running on flanged plate rails supported on stone blocks and travelled at about 8 km/h. The initial demonstrations were successful, hauling over 25 t of iron. However, the 5 t weight of the locomotive soon damaged the track and broke many of the cast iron rails. In addition, "*every movement of the engine caused a hideous snorting and clanking, accompanied by the loud noise of escaping steam*".[372] Although these set-backs might now seem relatively minor, the engine was subsequently converted to use in a stationary operation.[373]

In 1804 Trevithick also built a locomotive for a colliery at Newcastle upon Tyne. It was the first to use flanged wheels. It also proved too heavy for the intended rail track. For transport, there was a clear need for engines with better power-to-mass ratios.

Fig. 42 The first locomotive – produced by Trevithick in 1804 and operated at Pen-y-Darren in Wales. The horizontal piston is at the front. The very large wheel is a flywheel to maintain momentum between piston strokes. Note the iron L-section rails supported on blocks (see Fig. 41). The initial rails would have been made of short (1 to 2 m) lengths of cast iron. With permission of the Science Museum London (10301151).

Another early railway ran from Swansea to The Mumbles at Oystermouth in Wales. It was approved in 1804 and opened in 1806. The horse-drawn system was used to carry quarry products to Swansea and its claim to fame occurred in 1807 when it began to also carry fare-paying passengers. It was thus the first passenger railway although it operated more like a tram than train. It remained a horse-drawn service until 1877 when steam locomotives took over (Fig. 43). It has been estimated that in 1810, before steam locomotives were in use, Britain already had some 500 km of railway track in operation.[374]

Similarly, America's first commercial railway was a 5 km horse-drawn railway built in 1826 in Boston, Massachusetts, primarily to bring granite to the Bunker Hill Monument construction site. It used wooden rails with iron wearing-plates.

Trevithick's 1804 steam-powered railway locomotive attracted wide attention. He built a further locomotive at Bridgnorth (Shropshire) in 1808 and ran it briefly on a circular demonstration track in London before a broken rail led to the vehicle overturning. By 1823 there were 28 locomotives in operation, including three built by Trevithick. None were able to outperform horse-drawn haulage.[375] Trevithick died penniless in 1833. However, a talented new player was about to appear on the stage and change the railway scene for ever.

Fig. 43 The first passenger railway. The Swansea – Mumbles horse service began in 1807 although the drawing shows the service in 1870. via Wikipedia Commons.

George Stephenson[376] was born in 1781. He was uneducated and taught himself to read and write when, as a young man, he began working as a stoker for steam engines used by the local Northumberland mining industry. In his spare time he mended local clocks and watches. His mechanical skills and inventiveness did not go unnoticed and in about 1810 the Killingworth Colliery near Newcastle gave him the resources to develop new steam engines. In 1814 he produced his first steam locomotive, the *Blucher*, for use in the Colliery. The Blucher was of similar design to Blenkinsop/Murray rack-and-pinion locomotives developed in 1811-2 however Stephenson used flanged iron wheels running on Birkinshaw's new flat iron rails (Chapter 9c). The method relied totally on wheel-rail contact to provide the horizontal forces needed to cause motion. Within a year Stephenson's Killingworth locomotives had established steam haulage as a practical proposition and he was then widely recognised as an expert in steam engines and by many as a mechanical genius.

He had an exceptional forward view of how transport would develop. In 1821 at a dinner in Stockton, Stephenson told his son Robert and a young friend that "*railways will come to supersede almost all other methods of conveyance... railroads will become the great highway... it will be cheaper for a workingman to travel upon a railway than to travel on foot*".[377]

Exploiting the extensive coalfields at Darlington about 50 km south of Killingworth required the provision of a 40 km access link to wharves at Stockton on the navigable River Tees. The scheme was promoted by an entrepreneurial industrialist from the Darlington wool industry, Edward Pease, who could see how much improved accessibility would raise the value of the local collieries and so risked his significant fortune on the new scheme. The initial plans had been for a canal and then a horse-drawn railway. However Stephenson heightened Pease's risk when he proposed what would become the first practical steam-powered railway. The completed project quickly justified the expenditure and, in addition, was soon providing far more than local access to mines and wharves.

Stephenson undertook the design and construction of both the railway and the locomotives. The locomotive – *the Locomotion* – was based on his previous colliery locomotives and was a four-wheeled vehicle with two cylinders (Fig. 44). It weighed about 6 ton. The route began operating in 1825 and initially employed a combination of steam locomotives, steam-powered *winding engines* and horses. Winding engines were stationary engines which hauled wagons by pulling a cable attached to the wagon. This had been a common operation at mine sites and at the *inclined planes* used for hauling wagons and even canal barges up and down steep slopes. The opening of the Stockton – Darlington route and its subsequent success created great public interest (Fig. 45).

The historical relevance of the Stockton to Darlington railway is neatly summarised in a book produced for its 150th anniversary.

> *It was not the first railway as railways had been operating in mines and collieries since at least the 17th century; it was not the first public railway; it was not the first to use steam locomotives; and it was not the first to carry passengers. But it was the first to fuse them all together into a single, day-to-day working operation.*[378] [The earlier railways alluded to were discussed in Chapter 9c.]

Fig. 44 George Stephenson's *Locomotion* began operating in 1825. It was the first locomotive intended to operate on a railway track outside a colliery. Its two vertical cylinders are on the side away from the photo and can be seen operating two horizontal beams at the top of the photo. via Wikipedia Commons photo by Chris55 taken in the Darlington Railway Centre and Museum.

Fig. 45 Contemporary drawing of the opening of the Stockton to Darlington railway in 1825. From [379].

At Stockton, the locomotive option proved to be more reliable and of much lower cost than steam-powered winding engines. Horse-drawn passenger coaches also operated on the line until 1834 when they were replaced by the steam locomotives which had 30% lower operating costs. The British Parliament had had unwittingly hastened the development of steam power when it passed the Corn Laws in 1815 and forced a major rise in the cost of horse food.[380] As a result, it had become increasingly cheaper and easier to obtain coal to power locomotives than it was to find sufficient food for horses.

The Stockton railway attracted great national and international attention. An American engineer shrewdly commented "*There is no reason to expect any material improvement in the breed of horses in the future, while the man is not living who knows what the breed of locomotives is to place at command*".[381] In the year that the railway opened Carnot wrote in his monograph:[382]

> "*To take away from England her steam engines would be to take away at the same time her coal and iron. It would be to dry up all her sources of wealth, to ruin all on which her prosperity depends, in short, to annihilate that colossal power. The destruction of her navy, which she considers her strongest defence, would perhaps be less fatal.*"

Carnot, an engineer and physicist, was well placed to make such broader comments. His father Lazare was a member of the post-Revolutionary French government, his brother Hippolyte was a statesman and his nephew Marie Francois was to become Prime Minister of France.

CHAPTER 9E

STEAM ON THE ROAD

Trevithick's relatively poor experience with steam-powered road and rail vehicles has been discussed in preceding Chapters. This led him to concentrate after 1805 on the needs of industry for stationary power. Nevertheless, various other inventors and innovators took advantage of the improved manufacturing technologies and of improved steam engine design. Their focus was predominantly on the passenger market. In Britain their plans gained wide publicity and in 1817 the prospects for universal steam transport certainly seemed unbounded, as illustrated by the wonderful contemporary cartoon by Henry Alken in his "*The Progress of steam*" series (Fig. 46).

Fig. 46 Henry Alken's 1817 prophecy of the progress of steam-powered travel in 1829. The two prongs beneath the vehicle represented a common view wheels alone could not provide sufficient friction for horizontal traction.

Walter Hancock[383] was at the forefront of the new efforts. He had trained as a watchmaker and jeweller but became fascinated by the new steam engines. Working in Stratford in London in the 1820s, he introduced some major improvements to the existing engines to make them more appropriate for use in carriages intended to operate on roads and streets. Within the vehicle he used his vertical two-cylinder engines placed so that their pistons turned a transverse horizontal crankshaft. A chain drive was then used to transmit rotation of the shaft to the rear axle. The carriages required three operators: a driver, a boiler attendant and a stoker.

However, the major player in this steam-powered road-vehicle thrust came to be Goldsworthy Gurney,[384] who had trained as a doctor and became a prolific inventor. He had grown up in the same area of Cornwall as Trevithick and had seen and been influenced by his steam carriage. Gurney moved to London in 1820, captivated by the new steam technology. Over the next decade he introduced further improvements to steam engine design and operation. His horizontal two-cylinder double-acting engines operated at the relatively high pressure of about 1 MPa and produced about 8 kW of power. The pistons were directly attached to the rear axle by a crank. Concerns that the wheels would not develop sufficient traction on the road surfaces of the time were soon dispelled. The vehicles reached speeds of over 20 km/h, easily exceeding that of any horse-drawn alternative. Gurney's first vehicles weighed about 4 t, but he was soon able to reduce this to 2 t.

These first vehicles were effectively Landau coaches (Fig. 47) or stage coaches (Fig. 48) with the horses replaced by a locomotive.[385] Although not clear from the drawings, reports suggest that the engine drove only one of the two rear wheels.[386] The first public outing of a Gurney Landau vehicle was in 1829 when Gurney, investor Charles Dance and other supporters travelled in less than a day from the west of London to the city of Bath a distance of about 180 km. Speeds of 25 km/h were achieved on some portions of the trip, although frequent stops were needed to take on fuel and water. Gurney and Dance were attacked some 20 km before Bath by citizens of Melksham who viewed the new technology with fear and suspicion. The region had previously been a centre for Luddite and machine-breaker protests (Chapter 19b). Two years later an attempt by Gurney and Dance to open a similar service in nearby Gloucestershire was again disrupted by protesters, forcing the duo to pursue other interests (Chapter 13a).

Despite the set-back, the Duke of Wellington became a patron and used one of Gurney's machines to pull his Landau when he was at Hounslow Barracks in London's west. By late 1829 Gurney's steam-powered buses were competing at a practical and commercial level with the horse-drawn alternatives. There were lingering issues with respect to reliability and safety, but the Gurney vehicles had major operating advantages over horse-drawn vehicles and only stopped every 10 km or so to pick up fuel and water.

Fig. 47 Gurney's 1829 steam carriage at Hounslow. Note that two vehicles are used – one for the engine and the other a Landau coach for the passengers. This was intended to overcome the passengers' fear of the boiler exploding. The arrangement was commonly called a "*drag*". A horse-drawn Landau is shown in Fig. 11. With permission of the Science Museum London (10570452).

On his first commercial route – daily 12 km trips between Cheltenham and Gloucester – Gurney's operating costs were only 20% of those of the competitive coach service which required 18 horses per day per coach. Thus he was able to charge half the coach fare whilst completing the trip in half the time taken by the coach. Over time it became evident that not only could steam-powered vehicles haul larger loads at greater speeds than their horse-drawn competitors, but they also could be operated at about a third of their cost.[387] Despite this promise, Chapter 13a tells how Gurney's operations came to a sad and sorry end.

Gurney's initial success set the inventors' minds running and numerous steam coaches were tried in Britain over the next decade. Francis Maceroni[388] had worked with Gurney and then built his own single-unit steam carriage (Fig. 49) which ran successfully in both London and Paris.[389] Ex-navy captain Nathaniel Ogle[390] built carriages similar to Maceroni's. He ran services between Oxford and Birmingham (110 km) and between London and Brighton (90 km). His peak speed would appear to have been about 20 km/h, but Ogle publicly claimed much higher speeds. Such claims were probably necessary as in the 1830s the horse coach service on the London - Birmingham route – the "*Independent Tally-Ho*" – was routinely averaging 23 km/h. In addition, there was much verbal competition and bad-mouthing between the various rival inventors and operators and this is situation discussed in more detail in Chapter 13a.

Fig. 48 Gurney's single-unit steam carriage. It is said to have weighed 2 ton but it is not clear whether this vehicle was built and operated. The steering looks unrealistically fragile. No fireman or stoker is shown although in reality they were both needed at the rear of the vehicle. The confusion is typical of the claim and counter-claim that embroiled the industry at that time. An engineering drawing of the vehicle is given as Plate 13 in [391]. With permission of the Science Museum London (10570425).

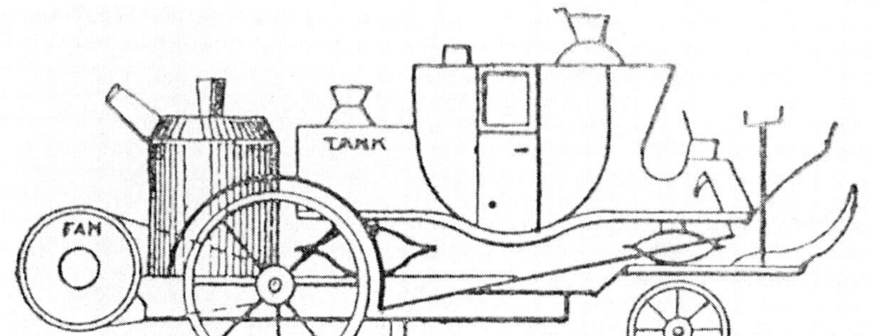

Fig. 49 Ogle's 1831 steam carriage. Maceroni's 1833 patent suggests a very similar arrangement. From [392].

Between 1833 and 1835 William James[393] designed, patented, built and operated a steam carriage which incorporated some major advances, such as three-speed variable gearing. A respected commentator notes that the James vehicle had "*many remarkable features*" in advance of other systems and which subsequently reappeared half a century later in the new internal-combustion vehicles.[394] James apparently found the exercise unrewarding and diverted his attention to railways and transport by pneumatic tubes, a technology explored a little further in Chapter 12c. Meanwhile Walter Hancock had been continuing his developmental work and in 1831 he operated a ten-seater steam carriage called "*The Infant*" over a 5 km route between Stratford and London. He used chain transmission which allowed him to provide spring suspension on both axles.

Chapter 12b will discuss how George Shillibeer operated his first horse-drawn bus service in London in 1829, after a Hackney carriage monopoly had been removed. The service ran between the inner City and Paddington and horse-drawn buses were to serve London well for the remainder of the century. Nevertheless, the market greatly tempted the steam carriage inventors. In 1833 Hancock produced the *Enterprise* steam-powered bus (Fig. 50) based on generally on Shillibeer's horse-drawn *Omnibus* (Fig. 68). The *Enterprise* was the first steam bus to offer regular services in London, running on the same route that Shillibeer used for his City to Paddington horse-drawn bus service.

Hancock soon became the leading steam carriage operator, with a fleet of nine steam vehicles providing services within London and to its suburban fringes. He also had a regular long-distance service between London and Brighton. Hancock operated profitably until 1840, when heavy road tolls, low profits and other factors to be discussed in Chapter 16 pushed the market back to horse-drawn buses. Hancock was bankrupted in 1845.[395]

Three major operational problems with all this generation of steam vehicles can be glimpsed in Figs 46-50. First, the high centre of mass and the distribution of much of the weight over the rear axle made them inherently difficult to steer. Second, the person who steered sat at the front of the vehicle and the stoker who controlled the power and the brakes worked at the rear. The two could not communicate effectively with each other. Finally, their large size made them difficult to operate in urban areas.

Fig. 50 Hancock's 1833 steam bus. It provided Britain's first scheduled bus service. Compare with Fig. 47 where the engine is in a separate vehicle, and note the improvement in passenger provisions over Ogle's coach (Fig. 49). Contemporary painting by William Summers.

The prospect of steam-powered road vehicles continued to attract many inventors and their efforts were carefully reviewed later in the century by Fletcher.[396] An 1831 House of Commons report stated that replacing animal power by steam power was "*one of the most important improvements in the means of inland communications ever produced*".[397] The impact was to be far more focussed than the House implied and the world entered a half-century when almost all transport development became focussed, not on road transport, but on the new steam-powered railways. This time of dramatic change will be explored in the following Chapters and we will return to steam-powered road vehicles in Chapter 13 when both the technology and the market place will be seen to have become very different to the scenario described above.

CHAPTER 9F

STEAM AT SEA

The practicality of the Boulton and Watt steam engines described in Chapter 8 attracted the attention of many inventors. An early maritime application of the steam engine occurred in 1796 when a rope-maker, John Grimshaw[398], installed a Boulton and Watt engine in a flat-bottomed boat in Sunderland (Durham). The engine was used to pull the ropes of a "*bag and spoon*" device which helped to dredge sea-bed material from shallow shipping channels.

A number of 18th century inventors had toyed with the concept of using steam engines to power ships, but with only modest success.[399] Nevertheless, engine size and weight were less of a concern for ship operators than for other transport modes operating in restricted spaces and so shipping provided the first useful applications of steam power to transport. John Fitch[400] had worked as a clock-maker and land surveyor. In 1790 he was taking fare-paying customers over 25 km of the Delaware River between Philadelphia and Burlington in the USA, using steam power to pull vertical paddles attached to an endless chain.

The first practical steam-powered boat (to be called "*steamers*") was the *Charlotte Dundas* which in 1803 pulled two barges along the Forth and Clyde canal serving Glasgow in Scotland. The pistons were mounted horizontally and turned a paddle wheel via a crankshaft. The steamer was built and operated by William Symington[401] who had been dabbling with steam-powered boats since 1788. He had honed his steam engine skills when working with Murdock (Chapter 9b) and had then produced an improved version of the Watt atmospheric engine at Dumfries, taking advantage of the recent expiration of the key Watt patent.

The Charlotte's journey was only a demonstration run and the first commercial steamer operation was due to Robert Fulton[402] who knew Symington and his work. He had tested a steamer of his own design on the Seine in Paris in the same year as the Charlotte Dundas demonstration.[403] These two important trials in 1803 did not help Henry Bell[404] who in the same year unsuccessfully tried to convince the British Admiralty to develop steamers. Only one Admiralty Lord – Horatio Nelson – supported his proposal. Napoléon was also unimpressed with steamers, allegedly commenting on reports of Fulton's boat "*How, sir, would you make a ship sail against the wind and currents by lighting a bonfire under her deck? I have not the time to listen to such nonsense.*"

In 1807 Fulton fitted a British Boulton and Watt engine to a ship called the *North River* (later renamed *Clermont,* Fig. 51). He used it to provide a steamer service on the Hudson River between Albany and New York, a distance of about 250 km. The average speed was 8 km/h. Steamers plying the large American rivers were to become a major component of the Industrial Revolution in that country, operating extensive passenger and freight services – a role later replicated by steam locomotives in the vast and relatively unoccupied American countryside. Fulton's success in America was not replicated in Europe until about 1812 when some steamers began operating in Scotland on the Clyde and the Firth of Forth.[405] The first was the *Comet*, produced and operated by Henry Bell. The Comet demonstrated to sceptics that a steamer could sail into a head wind, a feat that no sail boat could match.

Fig. 51 The first commercial steamer operation, New York to Albany, 1807. The ship is the *North River* (later *Clermont*). Project Gutenberg via Wikipedia Commons.

By 1814 steamers had proved their commercial worth in Scotland and England. From 1815 steamers were operating along the Thames in the vicinity of London, providing commuter services between Richmond and

Greenwich and day-trip services as far as the coastal resort of Margate – a journey that took about 6 hours.[406]

The next major initiatives came from David Napier[407] who realised that steamers would be far more effective if they had sharp prows. His 24 m *Rob Roy* was built in 1818 and its paddle wheels were powered by a 20 kW steam engine. It began operating across the Irish Sea between Glasgow and Belfast in 1818. In 1819, steamers were operating a regular service between Holyhead in Wales and Dublin in Ireland (Chapter 4c) and were being used on the Atlantic trade routes in the 1830s. In 1821 the *Rob Roy* was moved to Dover where it serviced the first steamer crossings of the English Channel. Coastal shipping and the first iron-hulled steamers came into service at about this time.

Steamers soon demonstrated enormous advantages over sailing ships in terms of manoeuvrability and ability to operate in all normal weather conditions. They also had a major advantage over all other modes in terms of load capacity. There were none of the carrying capacity limits faced by land vehicles and the forces resisting horizontal movement depended not on the mass of the cargo, but on the cross-section of the ship. Thus heavier cargoes were accommodated by simply the lengthening the ship. Screw propellers rather than paddle wheels had been invented before the steam era but a significant advance occurred in 1836 when Pettit Smith,[408] a farmer who built model boats, placed the propeller immediately in advance of the rudder.[409] Tests in 1839 convinced the British Admiralty of the overwhelming advantages of steamers with screw propellers. Henceforth, paddle wheels were rarely used. The main limit on the usefulness of propeller-based systems was the need to manage the longitudinal thrust developed between the propeller shaft and the body of the ship. This required very large and inefficient bearings and was not fully resolved until the invention of the Michell thrust bearing in 1907.

Coastal steamers had a major impact on the British economy with about 350 steamers in operation in the 1820s.[410] Various raw materials – particularly coal, tin, iron and copper ores – were moved routinely between producers and consumers and the reliability and frequency of the service meant that storage and warehousing costs and working capital needs were greatly reduced. It is noteworthy that these impacts preceded by some decades the subsequent impacts of steam railways.

As the “railway mania” to be discussed in Chapter 11b eased in the 1850s, investment capital began seeking new opportunities. The provision and operation of international shipping, taking advantage of steam power and the new iron plates caught the attention of British investors. The first of the new breed of ships was Isambard Kingdom Brunel’s[411] *Great*

Western. Brunel had already made a name for himself, initially in association with his father Marc, as an engineer with projects such as the first transport tunnel under the Thames in London (1825-1843, Chapter 5d), the Great Western Railway linking London to the west, and from Wales to Devon (1841-1844, Chapter 12a), the Royal Albert Bridge over the Tamar River between Saltash in Cornwall and Plymouth in Devon (1854-1859) and the Clifton Suspension Bridge over the Avon Gorge near Bristol (opened 1864 after Brunel's death).[412]

Brunel (Fig. 52) was unquestionably one of the 19th century's great engineers. His education was unique. Initially he was taught engineering and mathematics by his father, Marc, who was a French-educated civil engineer. Isambard was then sent to schools in England and France for a conventional "classics" education in literature and the arts. Finally, he was apprenticed to a watch-maker in Paris. Watch-makers were to play a major role in the Industrial Revolution (Chapters 14b&c).

The attractive shipping market was the Atlantic route between Europe and North America and Brunel's challenge was to use steam-power to capture that market. A salutary event had occurred in 1819 when an American-built sailing ship, the *Savannah*, with supplementary steam-powered paddle-wheels, crossed the Atlantic from Savannah to Liverpool carrying coal and ballast. Only 13% of the journey was made under steam and – to gain further cargo space – the ship was immediately converted back to a normal sailing ship when it returned to America.

Supported by the board of the Great Western Railway Company, Brunel set about designing and building a new steam-powered ship. He was greatly assisted in its design by the experimental and theoretical work of William Froude,[413] whose contributions to the behaviour of ships' hulls in water was without rival. Froude had trained in mathematics and then worked as a surveyor and engineer. In the 1830s Brunel and his colleagues realised that the power needed to push a ship through water depended on its cross-sectional area and not on its volumetric capacity. Thus the bigger ship, the more efficient it would be as a freight carrier. This concept drove the creation of the 1300 ton *Great Western* paddle-wheel steamship between 1836 and 1837. The design took full advantage of advances in steam-power and in the structural use of iron although the ship had an oak hull sheathed in copper. It used two steam engines with a total power output of about 500 kW. They were side-lever versions of the early beam engines (Chapter 8). One important advance was the use of condensers which allowed the steam used in the engines to be recycled, thus avoiding the need to carry large volumes of fresh water. The *Great Western* made sixty-seven Atlantic crossings over the next seven years and her best crossing took less than 13 days.

Fig. 52 Isambard Kingdom Brunel in front of ship's chains. William Powlett 1857 photo. With permission of the Science Museum London (10247110).

In 1840-1844 Brunel designed and built the *Great Britain*, which was then the world's biggest ship. It had a wrought iron hull made from iron produced at Coalbrookdale. It used propellers rather than paddle wheels and was powered by 750 kW steam engines with the pistons acting directly on the power shaft, rather than via a beam arrangement. The completed ship was too big to fit through the locks at the Bristol harbour entrance where it had been built and it took 18 months to resolve this access problem. The new design features introduced into the *Great Britain* meant that it could make the Atlantic crossing without refuelling. The

Great Britain is preserved today at the Great Western Dockyards, Bristol, England, in the same dry-dock that was used for her construction.

Before steam and cheap and plentiful iron, in the period between 1800 and 1840 inexhaustible supplies of good wood had helped America emerge as a major force in international shipping. However, the work of Brunel saw the advent of many British-built steamers which helped Britain to continue to dominate the international shipping market. Its share of world freight tonnage stayed at about 30% between 1840 and 1900, but its share of steam tonnage rose from 24% to 40%. During this time the total tonnage moved had increased tenfold.[414] Ship size also increased in a similar fashion. In 1838 the biggest ship using London docks had a gross weight of 2 kt; in 1900 the biggest was the *Celtic* at over 20 kt. In the 1880s the development of refrigeration meant that ships could carry meat and dairy products wherever markets demanded. By the end of the century steamers, which could ignore both wind and tide, had completely supplanted sail and were making the Atlantic crossing in 4 days.

As a consequence of the growing trade, there was major shipping congestion at Thameside docks early in the 19th century. One response was that merchants involved in the extensive British trade with the West Indies funded construction of a new West India Dock which used the Isle of Dogs and was accessed from the Thames downstream of London. It opened in 1802 and was a considerable advance on the earlier relatively unplanned docks. The growth in British trade was unabated and without parallel in other countries. Thus dock-building along the Thames continued for most of the 19th century.[415]

The massive development of ports during the Industrial Revolution is an often-ignored part of the Revolution. The ports were required to respond to both large increases in trade volumes and to quantum jumps in the size of the ships being handled. Of course, the direct economic value of the trade meant that, one way or another, the resources needed for such port expansion and improvement were always found. Two saving graces in this process were that the larger ships brought economies of scale in materials handling and that the new steam railways were able to service the docks without the need for large storage areas. All these changes greatly favoured riverside ports, such as London, Liverpool and Amsterdam, with ample dock areas, over many of the older ports which had developed predominantly as sheltered harbours.[416]

A major shipping break-through came in about 1860, at which time steam had only about 15% of the shipping market. It involved using high-pressure, compound steam engines in which the steam is expanded through a number of stages. Arthur Woolf[417] had developed such an engine in 1804, but its commercialisation needed the ability to make pressure vessels able to accommodate the necessary high pressures. With improved manufacturing technology, pressure vessel operating pressures did rise from about 100 kPa in 1804 to 700 kPa in 1860 to 1400 kPa in 1890. The major step forward in the 1860s occurred when John Elder[418], from the same firm as ship-designer David Napier, combined the new pressure vessel technology with his detailed understanding of Carnot's thermodynamic theory (Chapter 7b). His new engines reduced fuel consumption by a third and had a major impact on long-distance shipping.

By 1883 the tonnage of steam-powered vessels exceeded that of sailing vessels. However sail was cheaper to operate and so remained competitive on routes where time was not critical. For example, there were still 77 sailing ships carrying wool between Australia and England in 1891.[419]

Freight charges for international shipping dropped dramatically during the 19th century, sometimes to a quarter of earlier levels, particularly in the second half of the century. This was not only due to better ships but also to improved ports, scheduling taking advantage of the new telegraph services using undersea cables and improved navigation. Safety and security also reached a high standard. For example, the Cunard Line won its first trans-Atlantic mail contract in 1839. For the next 100 years, with the exception of the 1914-8 war years, it did not lose a single passenger or letter.[420] The Suez Canal (1854-69) greatly reduced travel times between Europe and the East and increased the effectiveness of steam-powered ships by permitting more en route refuelling stations. Construction of the Panama Canal commenced in 1881 but was not completed until 1914.

Although this book is about transport it is relevant to conclude this part of the transport theme by noting that steam power was rapidly being adopted by a whole range of activities. Nevertheless, transport generally was one of the more demanding of the applications as it imposed relatively stringent restrictions on an engine's size and weight. The general impact of steam power can be seen from the following usage data for the USA:[421]

year	Energy % from wood	Energy % from coal	Energy % from oil
1850	90	10	0
1910	10	80	10

PART 10

THE FIRST COMMERCIAL RAILWAY

CHAPTER 10A

SERVICING LIVERPOOL AND MANCHESTER

As described in Chapter 9d, over a short time in the 1820s the steam-powered railway had demonstrated its superiority for haulage to a previously sceptical commercial audience.[422] For many, a special attraction of the new railways was that they could provide a more reliable freight service than did the public road system. Not surprisingly, therefore, the next large opportunity for steam railways had a stronger and wider commercial orientation than either Black Country mines or the Darlington colliery.

The Industrial Revolution was already having a major impact on Manchester, close to both the Black Country and the great port of Liverpool. The population of Manchester had increased from about 25 000 in 1760 to 140 000 in 1824 as it became the dominant industrial centre. Indeed, it was the world's first "*industrial city*". By 1824 there were over 200 operating steam engines in Manchester and a thousand ton of goods were moving between Manchester and Liverpool every day. It was a two-way flow as the port of Liverpool was receiving almost 2000 bags of cotton per day from the USA.[423] Cotton was a critical cargo as for the entire 19th century cotton goods were the major British export item.

Until 1760 there was no road capable of carrying wheeled vehicles between Liverpool and Manchester.[424] Despite subsequent construction, the roads remained poor and the three canals were congested. Freight was hauled between the two centres on small and very slow horse-drawn wagons and canal barges. Port alternatives to Liverpool were not feasible as heavy road freight moved at an average of 3 km/h (or half walking speed) although packages that could be carried on the best horse-drawn wagons of the 1830s were delivered at about 10 km/h.[425] There were well-publicised cases where freight from Liverpool could reach New York in 21 days, which was less time than it had taken to travel from Manchester to Liverpool![426]

There was an obvious need for improved inter-city transport and it was readily argued that a Manchester to Liverpool railway would greatly enhance the value of local industry and of the seaport. In 1824 local businessmen formed a committee to progress the idea and one of its first acts was to appoint as its engineer George Stephenson, fresh from creating the Darlington railway (Chapter 9d). The enabling parliamentary Act was passed in 1825 with the strong support of large local manufacturers but in the face of vociferous opposition from groups affected by the proposed route, particularly land-owners, toll-road owners and operators, canal and coach operators, farmers and from the nobility who feared that their fox-hunting would be restricted.[427]

The money to build the railway was raised via public share issue. About 4000 shares were issued and over half were sold to citizens of Liverpool and Manchester. A major boost came when over a quarter of the shares were bought by the very wealthy Marquess of Stafford, who owned most of the canals in the region, including the highly profitable Bridgewater Canal which brought coal to Manchester.[428]

The promoters' decision to build the 56 km railway linking the two cities with twin parallel tracks from the outset underlined their confidence in their predictions of future traffic. Given the uncoordinated way in which road transport operated, an innovative decision was that all aspects of the railway – the track, the rolling stock and the servicing of customers – would be managed by a single entity.

The Liverpool and Manchester Railway Company, led by George Stephenson and his son Robert, won the bid to build the railway and work began in 1826. George was the engineer for the Line and Robert ran the firm that made the locomotives, although Robert was to be in South America until 1827. Construction was far from simple. For example, the track had to cross Chat Moss, a 9 km wide swamp, and the approach to Liverpool involved a 3 km long, 20 m deep, rock cutting through Olive Mountain and then a 2 km tunnel under the town to reach the docks. There were also 63 bridges, including a skewed bridge carrying the Liverpool – Manchester main road over the railway at Rainhill. Not surprisingly, progress was slower and more expensive than expected, and the Company had to be assisted with government loans from funds intended to help Britain recover from the Napoleonic Wars.

Based on common practice with rails at collieries, Stephenson had used *gauge* of 4 ft 8 inches for lateral rail spacing at Darlington (Chapter 9e) and had used the same gauge on the new railway. In about 1840 the gauge was increased to 4 ft 8.5 inches (1.435 m), mainly to accommodate locomotive behaviour on curves. This became the British standard gauge and remains today as the standard gauge in much of the world. Recall, by the way, that Chapter 9c described how the 1.4 m gauge dates from the animal-hauled wagons and was largely dictated by the width of wooden wagons and their haulage animals.[429]

The decision to operate the railway solely by the use of steam-powered locomotives was not made until late 1829. Indeed, at the time there was a common belief that horse- and cable-drawn systems were more effective than those which relied solely on locomotives.[430] Furthermore, George Stephenson had only achieved modest performance improvements since the 1825 *Locomotion* (Chapter 9d). Therefore the Directors of the Company organised a trial on a length of its new track at Rainhill to help them select the best locomotive. One key requirement was that the train had to be able to travel at an average speed of at least 17 km/h.

There were three serious entrants in the trials, which were watched by about 12 000 people. The Stephensons successfully offered the now-famous *Rocket*, which introduced many innovations over existing locomotives (Fig. 53). For example, the pistons were directly attached to the locomotive's two large front wheels using inclined cylinders. Using a large driving wheel meant that the vehicle travelled further on each piston stroke. The engine introduced the multi-tube boiler where the hot gases from the fire-box were taken through the water boiler by bundles of relatively small copper tubes, thus greatly improving the heat transfer process. The multi-tube boiler had first been used in limited form by Marc Seguin in Paris in 1828. At the same time Seguin was also making major contributions to the design of suspension bridges (Chapter 5c). In later years, Robert Stephenson would make many further improvements to locomotive design.

As a consequence of such innovations the *Rocket* had a top speed of 45 km/h and readily operated at 40 km/h. This must have been a great shock to those who had given prominent evidence to the preceding Parliamentary enquiry arguing that locomotive speeds could not exceed 7 km/h.[431]. It was also a major transport watershed, as it was the first instance when vehicles could travel on land at speeds consistently well in excess of those provided by animal-drawn transport.[432]

As discussed in Chapter 9e, Goldsworthy Gurney had steam carriages which could have been readily modified to take part in the Rainhill trial and challenge the *Rocket*, however he appears to have ignored the challenge.

Fig. 53 Stephensons' Rocket, in the London Science Museum. With permission of the Science Museum London (10680902).

The Rocket and similar engines were mainly used for hauling passenger trains. For freight haulage late in 1830 Robert Stephenson introduced the four-wheeled *Planet* locomotive and it became the model for many subsequent locomotives (Fig. 54). Rocking of the locomotive was reduced by replacing the Rocket's sloping pistons with horizontal ones which directly drove cranked axles. The *Planet* was the ninth locomotive built for the Liverpool-Manchester line. It was also the first locomotive design to be built in large numbers. The cranked axles proved an unsuccessful innovation as they frequently fractured. A further necessary modification was to increase the length of the boiler – and thus of the locomotive – to allow the furnace heat to be more widely distributed and the peak temperatures to be reduced. However this change reduced the ability of the locomotives to negotiate sharp curves in the rail track.

Fig. 54 Engraving of Stephenson's Planet near Edgehill Station, Liverpool, in 1831. A restored version of the locomotive is kept at the Science Museum in Manchester. Note that the structural supporting frame is now outside the wheels and the pistons are between the wheels. With permission of the Science Museum London (10419593).

The *Rocket* and the *Planet* highlight another major change that arose during the first half of the 19th century. It is helpful to realise that the typical travel speeds of the Rocket and the Planet of 40 km/h can also be expressed as (about) 10 m/s. With the circumference of the driven wheels in Figs 49 and 50 being (about) 5 m, to travel 10 m in a second requires two revolutions of the wheel – and of the piston system – every second, or about 2 Hz. This would mean that the spokes would only be seen as a blur. More generally, it meant that machines were now being built and operated beyond normal perceptions. For the first time inventors and builders had created machines that worked in a manner that they had conceptualised but could not have visualised. This was truly a new world.

CHAPTER 10B

THE TECHNICAL IMPACT OF THE ROCKET

When the Liverpool-Manchester railway opened in 1830 it was the first railway in the world to rely totally on steam power and the first inter-city railway. From the outset, steam-powered travel was seen as a commercial venture, both by the immediate service providers, such as Gurney and the Stephensons, and by the capital providers who invested in the new services. It was also evident from the beginning that the new railways, unlike improvements to the road system, would not be created by minor incremental changes. They would impose a completely new form on all aspects of society – from rural landscapes to time-keeping to urban stock markets. Why was this? There were five major factors contributing to a quantum change in transport provision.

First, a key feature of the new railway system was that, in order to carry the heavy locomotives, it used vehicles with iron wheels running on iron rails. Both wheels and rails were relatively smooth and provided a much lower coefficient of friction than would the then-conventional road vehicles running on conventional roads. As a consequence, lower horizontal forces were available for traction and braking than would be the case with rougher surfaces. This had initially caused some scepticism about the practicality of a system relying on iron wheels running on iron rails. The effect was partly countered by the fact that the vertical loads were very high and the contact areas relatively small, producing high contact pressures. As a result traction and braking were generally adequate, particularly when the wheel rotation was only gradually changed. However, rapid accelerations and sudden stopping were not feasible. Stopping in particular required early awareness of the need to stop and this required good sight distances, with no sharp crests, no tight curves and the development of a signalling system providing advance notice that stopping would soon be required.

There were no signals on the Stockton-Darlington line (Chapter 9d) when it opened in 1825 and none on the Liverpool-Manchester line (LMR) when it opened in 1830. In 1833 the LMR began using men standing beside the track to employ arm signals to convey stop/go messages to locomotive drivers. The arm signals soon came to be supplemented by flags and discs.[433] The first lit signal was introduced in 1834 and was simply a lamp on top of a post and was operated by a person climbing a ladder and changing the light to either white or red. The use of red for danger had been taken from shipping practice.[434]

These technical issues combined with a desire to reduce decision-making by drivers resulted in a system that was relatively ponderous, awkward and space-consuming. As rail systems became more intricate, very complex arrangements of mechanically operated semaphore arms and rail intersection *point* controls gradually developed, as illustrated by the late 1860s example in Fig.55.

The second technical issue causing major change was that the requirements for vertical alignment were severe due to both the smoothness of the rail-wheel interfaces and to the extra power needed when operating on grades. In such cases the extra traction required to meet the gravitational component of a long string of hauled wagons could only be supplied by the wheels of the locomotive. Tests by the Stephensons had shown that grades steeper than 1 in 200 were sub-optimal and Liverpool-Manchester route was built to a grade of about 1 in 2000 with slightly steeper grades at each end. For comparison, the desirable maximum road grade for the Royal Mail coaches had been about 1 in 20 (based on pulling vehicles up-hill) and maximum modern road grades are about 1 in 10 (based on poor sight-distance when travelling fast[435]).

The value flowing from the creation of a firm and flat alignment cannot be underestimated. Once a train has reached a steady, constant speed the only forces to be overcome are those due to wind and rolling resistance. Economies of scale become very real. The weight of the locomotive and of the cargo being hauled become almost irrelevant.

A third technical factor was that steering relied mainly on forces between the wheel flanges and the rails and was of limited value due to the centrifugal forces which increased with speed squared. Consequently, railway routes had to be located using horizontal curves with radii of over a kilometre. Even at low speeds, the almost right-angle turns of road vehicles were not possible.

As a consequence of these factors, the alignment of a new railway was far more difficult to accommodate than that of a new piece of road. The severe limits on vertical and horizontal alignment continued in the years ahead and meant that the location of new railways often caused major local disruptions. Indeed the euphoria that surrounded many of the new railways meant that they also blithely blocked many roads and streets which had previously provided local access.[436]

These far stricter restrictions on alignment were of their own sufficient to make the new railways very different to the road networks they were

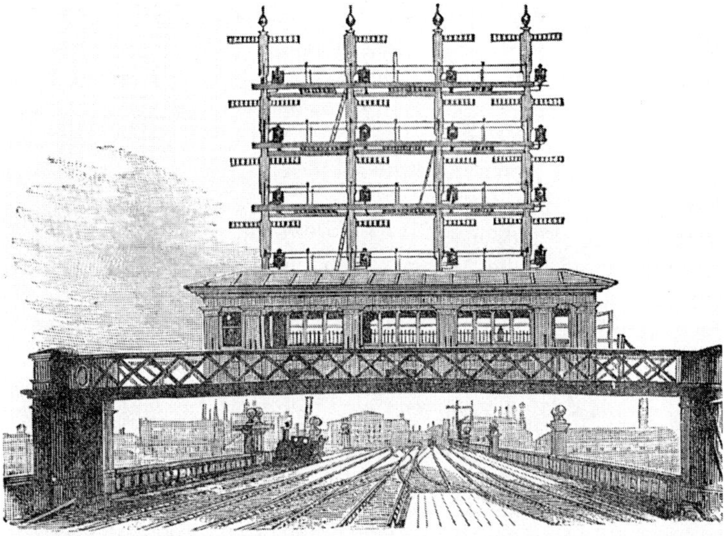

CANNON STREET STATION SIGNALS.

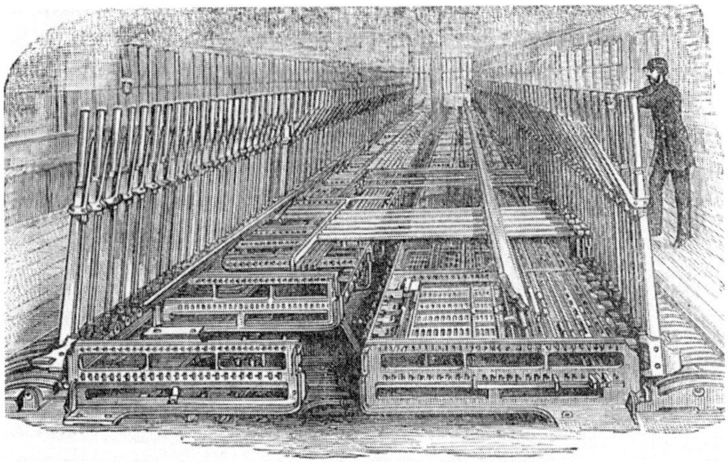

INTERIOR OF LONDON BRIDGE SIGNAL-BOX.

Fig. 55 The signal system at Cannon Street Station and its London Bridge approach in London in the late 1860s. The levers controlled the semaphores and the rail intersection "*points*". From [437].

supplanting. However, there was one other, fourth, major geometric restriction which arose from the fact that humans had evolved, slowly, in a low speed environment (Chapter 1a). People can see objects of about 2 m in size at a distance of about 1.5 km, but have a very poor ability to assess the speed of that object if it is coming towards them.[438] Helped by binocular vision, humans have a reasonable ability to judge distance, but judging speed involves perceiving rapid changes in that distance. Thus, for most of the human evolutionary period, any object that was detected in the distance would take at least 15 minutes to reach the observer. There was no need to instantly step aside. However George Stephenson's Rocket had a quantum jump in speed and could reach an observer within just 2 minutes of being sighted. Stepping aside needed to be done in a much sprightlier manner.

Proof of this came on the *Rocket*'s first official journey in September 1830, pulling a long string of wagons loaded with passengers and coal and an ornate coach. The site was the Rainhill Level portion of the new Liverpool-Manchester railway. The official welcoming party included the Duke of Wellington as Prime Minister and William Huskisson, Minister for Trade, a strong advocate for the railway and widely expected to be the next Prime Minister. The Duke was in the ornate carriage and Huskisson was in a less ornate separate carriage. Their train was on the southern track and had stopped at Parkside to allow passengers to wave as the *Rocket* passed by on the northern track. Although the *Rocket* could be seen approaching in the distance, Huskisson left his carriage and walked along and across the *Rocket*'s track to congratulate the Duke. Alas, the *Rocket* arrived much sooner than Huskisson had estimated and the erstwhile Prime Minister was fatally struck by the speeding *Rocket*-hauled train.[439]

As one consequence of this event, the British parliament decided that all railways should be fenced with no access other than at legally defined and well-separated locations. Abutting properties had no access rights, in strong contradistinction to the universal situation on public roads. This total restriction on side access had three key advantages for the railway operators: trains did not have to slow or stop for unplanned cross or merging traffic; they did not have to slow down for random crossings by people and animals; and revenue could easily be collected from users before joining trains at stations and sidings.

The fifth factor causing quantum change was that the commercial impact of the railways was not at the margin – rather, they dramatically altered the cost and convenience of transport and thus caused major commercial impacts. These are explored in the next Chapter.

PART 11

THE CONSEQUENCES OF THE NEW RAILWAYS

CHAPTER 11A

COMPETITIVE ADVANTAGES

In 1830 there were over 30 horse-drawn passenger coach services operating each way between Liverpool and Manchester at speeds of about 20 km/h and annually carrying about 100 000 passengers. In the same year George Stephenson opened his Liverpool-Manchester railway service. Not only was railway travel half the price of coach travel but, as described in Chapter 10b, it also offered greater reliability, comfort and freight capacity and all of these at speeds of up to 45 km/h. Within two years only one horse-drawn coach remained in operation. The impact on canal traffic was not as severe as, although an order of magnitude slower than the new trains, canal barges provided a much cheaper service.

The Liverpool-Manchester service was the first railway to use "*all the essential elements of the modern railway*".[440] As noted in the preceding Chapter, inevitably a new set of operational problems arose. Nevertheless, the railway was immediately profitable, carrying over a thousand passengers per day and 460 000 in its first year. This patronage provided investors with a 9.5% annual return on their capital. It also demonstrated that the railway not only met existing demands, but had created a whole new travel market.

The financial success of the Liverpool-Manchester railway quickly inspired investors to fund the construction of the 130 km Grand Junction line from Warrington, on the Liverpool-Manchester route, south to Birmingham. That line was approved in 1833, built on a route selected by George Stephenson, and opened in 1837. Growing confidence with steam locomotives had led to it being built with steeper grades than earlier routes. It was the first of the trunk railways and marked a further step in the capture of the markets previously served by the coach and wagon network. Subsequently, the financial success of the new railway companies allowed them to buy out coaching companies which were only indirect competitors.

In a period of remarkable activity, by 1838 a privately funded railway service was also operating between London and Birmingham. It was the beginning of the railways operating as a network rather than as a collection of point-to-point services. The 200 km route had been approved in 1833 and built by Robert Stephenson to a grade of no more than 1 in 330. It consequently incorporated many major bridges, tunnels and cuttings. The London end of the route took passengers to Euston Station on the then outskirts of north-west London and freight to a nearby terminal on Regent's Canal near Primrose Hill (see the later Figs 59 & 60). The canal provided a cargo link to the London docks which had proved impossible to provide by railways due to the high cost of the intervening inner-urban land. This inner-city land issue is discussed further in Chapter 12a. By 1846 a single company was operating a service between London and Manchester.[441]

In 1839 the British Post Office switched its London-Dublin service from road between London and Holyhead and then steamers to Dublin, to railway from London to Liverpool and steamers (Chapter 9e) from Liverpool to Dublin. The operators of the competing horse-drawn coach services were quickly forced to close by the market success of the superior new product.

In 1844 work began on linking the London – Chester – Liverpool railway service to Holyhead by a new line from Chester to Holyhead. Robert Stephenson was the engineer for the project and he used a route surveyed by his father George. In 1850 the connection was completed and a steamer operated the Holyhead-Dublin link.[442]

The travel times between London and Holyhead are indicative of the changes that had been occurring. In the coaching days in the 1820s, the trip took at least 41 h. After Telford completed his road-works in 1830 (Chapter 4c), trip times dropped to 28 h. In 1861, two trains carrying a vital diplomatic message completed the journey in a few minutes under 5 h.[443]

When the Great Irish Road (Chapter 4c) was approved by the British Parliament it had been intended that the road service from London to Holyhead would be operated by steam carriages and Thomas Telford and Sir Henry Parnell had actively promoted their operation by creating a new steam carriage company. The railway investors saw them as major competitors and little quarter was given. Richard Creed, Secretary of the London to Birmingham Railway Company predicted publicly that much of any investment in the proposed Holyhead Steam Coach Company would be lost. He stated that "*The old race of Civil Engineers fear that the railways are destroying their fame and fortune*".[444] The railway lobby further argued that running steam carriages on common roads would be "*fraught with danger to the lives and property of His Majesty's subjects*".

Telford had died in 1834 and so did not see the consequent dramatic decline in the usefulness of his Great Irish Road. He had been openly sceptical about using steam-powered locomotives on railway tracks. He thought that the exclusive railway track limited the services that could be provided and gave its operators a service monopoly. Like Daimler (Chapter 14b), he sensed that there must be something better than the steam-powered railway locomotive. Telford's view of the future saw much improved steam-powered road vehicles operating competitively on publically-owned roads. For the first 40 years after his death his vision seemed wrong. And then along came the internal-combustion engine!

The first railways and canals were constructed by hand and by the use of horse-drawn carts. A major change occurred following William Otis'[445] 1835 invention of the steam-powered mechanical excavator (or shovel). Most of the initial batch of seven built in Philadelphia were used on railway and canal construction. By 1842 an Otis excavator was working on railway construction in Essex in England.[446]

CHAPTER 11B

INVESTMENT "MANIA"

Following the success of the first railways, investors' eyes glinted at the prospect of more steam-powered railways and a boom then occurred in funding railway construction. The enthusiasm for railways was unbridled and between 1825 and 1837 the British parliament passed over 54 Railway Acts which authorised the construction of 107 privately owned railways operating about 800 km of track with at least a further 1000 km legally sanctioned.

In 1843 Britain was emerging from a period of depressed trade. Nevertheless, some 3400 km of railway routes were then in operation and it has been observed that "*by 1843 England stood on the threshold of the great railway mania*".[447] The term "*railway mania*" was well-used then and in subsequent decades. Related terms such as "*Railway to Heaven*" and "*Railroad to Hell*" also became common place.[448] By 1847 about 18 000 km of British railway routes had been sanctioned.[449] The investment bubble then burst and many investors suffered catastrophic losses in the final years of the 1840s, although most of the railway companies continued to operate. Another financial collapse occurred in the 1860s when British railway companies over-invested in the provision of railway facilities in other countries.

Some have suggested that the "*great mania*" had been preceded by two earlier manic periods. The first was in 1824-26, when the Stockton to Darlington railway had hinted at the potential of railways (Chapter 9d) and the second in 1831-37 when the Liverpool-Manchester railway (Chapter 11a) had demonstrated that potential.[450]

The high returns from the first railways were certainly a major cause for the railway mania. An important secondary cause was that investors realised the importance of claiming and monopolising new geographic markets and that this could only be achieved by being the first to build a railway into that market region. Thus parliamentary approval of a proposed route, successful fund raising and speedy construction were of paramount importance.

The British Government's role in the process was to formally approve or reject each proposal – most importantly granting the power of compulsory land acquisition to an approved project. The proposals had to give precise details of their proposed alignments and this led to intense political debate at local, regional and national levels. Many deals were done to achieve approval. However there is no evidence that schemes were chosen to fit some broad set of national transport objectives.

In 1842 when the core of the British railway network was in place, passenger traffic provided 70% of the railways earnings.[451] Nevertheless, freight was seen to be the main future revenue stream for the railways. Before railways, some British canal companies – which relied almost totally on freight revenue – were paying dividends of over 25%. With the coming of steam railways there was now major competition between influential groups of financiers and their vested interests in either the new steam-powered railways, or the older canals, or the toll roads serviced by the new steam carriages.[452] The railway companies dominated due to the much better potential returns offered by their facilities. And they often did more than just dominate as they bought canals to such an extent that by the 1870s the majority of British canals were owned and operated by railway companies. The canal companies had often been short-sighted, spending their money on opposing new railway proposals rather than on improving the services that they were offering.[453]. The three main motives that the railway companies had for canal acquisition were to stifle vociferous objectors to new railway proposals, to acquire useful land for new railway routes and to diminish potentially competitive services by either closing the canals or greatly increased the fees that they charged canal users.[454]

In summary, the commercial advantages of railways to investors were that:

1. Railways could be located to service routes where there would be a known high demand for their services.
2. Railways were able to carry large volumes of high-value freight and large numbers of affluent passengers.
3. Paying for the railway transport service was seen by users as appropriate.
4. Joining and leaving the trains could be closely controlled, to both minimise stops and maximise fare collection.
5. Astute (and often forewarned) investors could easily capture a large share of the land-value increases that flowed on from the enhanced accessibility provided by the new railway.
6. The railway companies owned both the rail track and the vehicles that operated on the system. They thus had a complete monopoly

over a route and their revenue source was far wider than that of a toll road company. Indeed, although both privately-operated toll roads and canals were required to accept any appropriate toll-paying user, the railway companies successfully argued from the outset that scheduling and operational problems meant that they had to have complete control over who could use their facility. Indeed, it was common for the owner of a railway to also be the sole operator on that railway.[455]

In Britain this was a time when there was plenty of available capital. Moreover, that capital was actively seeking railway investment, goaded by the returns of ten per cent or more being offered by the first couple of railways. Money was cheap and politics were calm. In the course of one month in 1845 some 357 speculative railway projects were advertised, requiring an aggregate capital of 350 million pounds.[456] The largest single scheme approved by the process was the Great Northern line from London to Doncaster.

One relevant summary is worth quoting "*Investors were prepared to plunge wildly into a rising market, regarding every new proposal as a stereotype of the best*".[457] However, the initial projects were schemes with special advantages. Later projects became increasingly marginal and as time passed many of the newer ones did not provide a useful return on investment.

The railways were planned, built and operated by a disparate range of private companies. Unlike the turnpike trusts which had had to raise their capital by debts with defined interest payments, the railway companies were allowed to raise capital by issuing shares, with no firm promise of future dividends or capital appreciation. Many were to subsequently fail as their revenue failed to cover their debt servicing and operating costs. Their demise was postponed by the unhealthy practice of allowing the railway companies to pay dividends out of their relatively large capital raisings.

The fact that a useful and coherent operating system did arise was due mainly to the incidental efforts between 1835 and 1849 of George Hudson,[458] known as the "*Railway King*". During two days in 1846 he gained parliamentary approval for 40 railway schemes.[459] However, later in the year his huge railway empire collapsed due to unsustainable debts and Hudson's personal acts of dishonesty, fraud and bribery.[460] Nevertheless, by the end of the century there were few towns in Britain that did not have at least one railway service, and most of the surviving services were providing a reasonable return to investors.

CHAPTER 11C

RAILWAYS BEYOND BRITAIN

The railway invention was picked up rapidly around the world and the investment potential of steam-powered railways quickly captured financiers' attentions in most major countries. In 1828 railways were operating between Budweis in the Czech Republic and Lintz in Austria and between St Etienne/Andrézieux and Roanne in France. Elsewhere railways began operating in the USA between Baltimore and Ohio in 1829, in Spain between Reusse and Tarragona in 1830 and in Belgium between Brussels and Mechelin in 1831. The French project was undertaken by the Seguin brothers, well-known for their innovative suspension bridges (Chapter 5c). Marc Seguin had also been active in steam engine development (Chapter 10a) but the brothers brought their locomotives from the Stephensons.

Major investment in European railways followed the British experience but with a time lag of a decade or so. Many of the facilities were supplied by British railway interests and particularly by Robert Stephenson. In Germany and Austria, major railway networks were built during the 1840s, despite the fluid nature of the distribution of political power between the then independent states. The railways provided a major boost to German industrialisation. Contrary to the British laissez-faire practice described Chapter 11b, central planning of the railway network occurred under government control in France (from 1842) and Belgium[461] and, indeed, in most of Europe. On the other hand, in the US government policy was to offer incentives for railways to be built by investors rewarded by the opportunity to develop the vast countryside.

One of the major European transport impediments was the crossing of the Alps and Napoléon's improvement of the road crossing was discussed in Chapter 3b. The Swiss and the Austrians soon built competitive trans-Alpine railways. However, there remained a need for a crossing too far west to be controlled by the Austrians, but still able to capture the sea traffic using the new Suez Canal (Chapter 9f). The resulting demand led to the construction of the Mont Cenis Rail Tunnel. Major work started in 1861 at the end of the Italian war of liberation which had led to the ceding of Savoy and the eastern part of the Pass to France. The 14 km long, dual track tunnel from Modane in France to Bardonecchia in Italy opened in 1871 after more than a decade of construction. It replaced horse-drawn mail service (Fig. 17) which in winter could take 15 h to traverse the Pass.

As discussed Chapter 11b, British investment in British railways fell markedly after 1850. Consequently, British capital became heavily involved in the development of European railways in the 1850s. However, there were other competitive investors entering that market so much of the resulting "spare" British capital in the latter part of the century was then invested in the growing American railway system. This led to the same total diversion of capital and human energy from roads to railways in America as had occurred previously in Britain.

The Baltimore and Ohio railroad opened in 1829 and was envisaged as a common carrier for freight and passengers, following the model of the Surrey Iron Railway (Chapter 9c). Initially, it operated with horse-drawn vehicles. The first steam-powered locomotive to operate in the US was built in the West Point Foundry in New York and initially operated in 1830 on 10 km of track in Charleston, North Carolina, prior to successfully serving the needs of the local cotton exporters.[462]

Some British locomotives were tried on the early American railroads but they did not operate well on the American tracks which were built to much cheaper and more affordable standards for grade, alignment, roughness and road and river crossings. In this context the US found it more cost-effective to develop its own robust, practical and powerful locomotives and equally robust railway and track systems (Chapter 9c). For example, the '*cow-catcher*' at the front of a locomotive was added as most of the US railway system was unfenced.[463]

By 1850 America, with its vast countryside ripe for development, had 15 000 km of railway, by far the world's greatest system in terms of length and already 50% greater than English network. The American Civil War had broken out in 1861 and in 1862 President Abraham Lincoln signed the Pacific Railroad Bill, mainly to ensure that the West was strongly aligned to the Northern forces of the shaky Union. Consequently, a major event in railway history occurred in 1869 when America's transcontinental railway came into operation.

In the extensive and relatively prosperous expanses of the United States, the railways were a dominant commercial force at the end of the century. Their major role in the latter part of the 19th century has been well reviewed by Gordon.[464]

American railways were not always restrained in exercising their power. In a review of that time, Trachtenberg noted "*railroad corporations were seen as the epitome of ruthless, irresponsible business power, a grave threat to order and stability*".[465] Politicians such as the silver-tongued William Jennings Bryan took advantage of this view attacking the "*public be damned*" attitude of the large railroads who were cast as arrogant, impersonal, unsavoury, land-grabbing and opportunistic. David Belasco, a well-known theatrical producer observed at the turn of the century, that "*the train had gone from representing the path to a glorious future, to symbolising the all too alienating present*". The situation was exacerbated when America entered a major economic depression in 1893. A prime cause of investor panic at the time was speculation about the capitalisation levels of the railroad companies. Within two years, the entire US railway network was on the brink of collapse and, with the coming of the car, never regained its earlier standing.[466]

In the 1890s investment in European railways was also no longer an attractive proposition. However, on a wider international level, the momentum for railway investment was still strong as the following Table illustrates:[467]

Year	**World railway length Mm**
1840	0
1850	40
1860	110
1870	220
1880	380
1890	630
1900	820

The three largest systems were now USA (280 Mm), Germany (50 Mm) and France (40 Mm).

CHAPTER 11D

RAILWAYS' IMPACT ON ROADS

As described in Chapter 11b, British investment in the new railways boomed in the 1830s and 1840s. This left virtually no public or private capital for roads, other than those which provided access to the new railways. For the toll road companies, the railways brought the additional disadvantage of attracting travellers with the greatest willingness and ability to pay tolls. This would typically include hauliers of valuable freight and well-paid long-distance travellers. The remaining possible toll road customers were local travellers and local farmers who had little or no ability to pay the tolls needed to even maintain major roads, let alone return a dividend to investors.

Consequently, key roads were often over-priced and poorly maintained. Many potential travellers became increasingly disgruntled, with a range of negative reactions. They reduced the number of trips that they made, they avoided the toll booths, they evaded the toll collectors and they protested loudly, disrupting any remaining travellers and destroying toll-collecting facilities. The best known of these protestors were the Rebecca Rioters in South Wales in 1843. The rioters disguised themselves as women, to avoid identification. "*Rebecca*" was a biblical reference.[468] They operated at night destroying toll booths and causing the death of at least one toll collector (Fig. 56). When captured and brought to trial, many were acquitted by sympathetic local juries.[469]

Fig. 56 Rebecca Rioters protesting about a toll-gate in South Wales in 1842. From [470].

Progressively, the various toll road trusts collapsed through real or effective bankruptcy. As a result, in 1864 the British parliament decided not to renew the rights of any turnpike trust when its existing term expired. Ironically, the last operating toll road on a major road link was on the Anglesey portion of the Great Irish Road (Chapter 4c). Its terms were not renewed in 1895. A few toll facilities survived the measure as they were privately owned and on private roads.

Existing highways fell into disrepair. At the peak of the coaching era many major roads, such as the Great North Road from London to Edinburgh via York and Durham, were 30 m wide which is comparable to a modern freeway.[471] However, as traffic diminished only one lane in each direction was required. This needed no more than 7 m and so grass, vegetation and adjoining land-owners soon claimed the unused land. By 1885 many highways were described "*as looking like country lanes*".[472]

The consequential impact of rail on roads was mirrored in most countries. As an example, my home town of Melbourne in Australia had only been settled by Europeans in 1835 and its first railway opened in 1854, running from the new town to its port. For the next 60 or so years, transport resources were almost entirely devoted to rail. The railways were built and operated by the private sector. When, one-by-one, the operators went bankrupt, the government took over the provision of the service.

Between 1876 and 1894, annual expenditure on roads in Melbourne averaged only 3% of the equivalent expenditure on railways.[473]

Indeed, a major negative consequence of the new railways around the world was that attention and, more particularly, resources were diverted away from the road networks which did not merely fail to keep pace with transport demands but fell into disrepair. There was no vision of the suite of measures needed to provide a functioning total transport system. The scene was well painted by Schreiber, a German commentator, who wrote:[474]

> *The (latter part) of the 19th century was the age of the railway. That was inevitable. What is surprising, however, is that no one appears to have looked beyond the railway. It was, of course, a new and even unexpected development, but even level-headed and open-minded people regarded it as an end in itself, the perfect end-point which reduced to absurdity all attempts to improve on it. People were obsessed by the railway, lived for it, and built their politics around it .... In 1913 the road was never mentioned, except to remark in passing that some projected railway line or other "will replace a motor-road at present in existence": a grotesque view which shows how few statesmen can see even three decades ahead.*

Schreiber's criticism may be a little harsh as it seems clear from many aspects of the review reported in this book that people simply do not possess the ability to see 30 years ahead in a time of major technological changes. Therefore an alternative to Schreiber's criticism is that the absence of such prescience should have been recognised as the normal human condition in both operational decisions and in forward planning.

In addition to the competitive advantages of rail, there were a number of additional reasons for the collapse of the toll road system. The collapse is often painted in a political context, but – as constituted – the toll operators were doomed to fail no matter what the political climate might have been. One problem was that the location of the toll gates was dictated by the administrative structure of the toll company and took no account of the needs or purposes of travellers. Thus it was easy for travellers to perceive the system as unfair and inappropriate.

A second problem was that as vehicle technology slowly improved and vehicles travelled faster, stopping to pay a toll become even more irksome and it became increasingly possible to speed past or around a toll booth without being stopped to pay a toll. Further, as the road system became more interconnected it became easier to find alternative cheaper or toll-free routes. Finally, the tolls collected had to pay the toll collectors, fund road maintenance, and meet interest and debt repayment charges. The task became steadily more difficult.

A more comprehensive system of collecting revenue from road users was needed and, as self-powered vehicles became more common, fuel taxes were quickly instituted by Treasurers around the world. They were easy to apply as they could not be avoided, were collected by the private sector as fuel sellers and then passed quietly onto the local Treasury. They could be given a positive face by painting them as a form of consumption tax. Indeed, in 1907, Herbert Asquith the Chancellor of the Exchequer, told the British Parliament that a tax on cars was an ideal tax because the car was a luxury that was apt to degenerate into a nuisance.[475] A year later a Finance Act imposed just such a tax on British road users.

It must emphasised that the advent of railways did not result in the end of horse-drawn travel, even in countries such as England with intense railway networks. Many towns were still not serviced by rail and rail did not serve all the routes between towns. The steam-powered railways and canals could only provide services for passenger or freight movement if the demand for those services was high. Furthermore, passengers and freight still had to be brought to and from railway stations and sidings. Finally, many journeys did not involve towns as either origins or destinations. So all those local and infill trips made today by car or truck still had to be made on foot or by horse-drawn vehicle.

Apart from the bicycle (Chapter 2) and the late development by the French of light steam-powered cars (Chapter 13b), in these circumstances the only alternative to animal-powered transport was the ungainly steam carriage whose use was severely hampered on major roads by restrictive regulations and in the countryside by the poor alignment and condition of the road network (Chapter 13b). Thus, as discussed above, the supply of rural road space was actually falling. In the cities, personal transport was predominantly by horse-drawn bus and tram, although towards the end of the century electric trams (Chapter 12c) and underground railways (Chapter 12d) had begun serving some of the urban market.

Consequently passenger and freight transport still relied heavily on the horse and the ox. The streets of cities and towns were very much the province of horses and pedestrians. Indeed, in England the introduction of the railways actually increased the demand for horses and in the time of railway ascendancy from 1840 to 1870 the number of privately-owned large horse-drawn carriages grew four-fold from 30 000 to 120 000 and the number of light carriages grew six-fold from 40 000 to 250 000.[476] In 1900 America had one horse for every four people. It is estimated that London had about 180 000 working horses at the end of the century.[477]

In addition, careful horse breeding practices, better harnessing and lighter vehicles all meant that the 19th century horse was an increasingly useful mobile power source and was in high demand. However, it had many drawbacks some of which have already been touched on in Chapters 1e and 13b. Coach services required two new horses for every three kilometres of travel and it took about ten horses to keep one urban bus in service. Each horse ate as much eight adult humans and horses consumed about a third of the crops produced on US farms. It was estimated that in New York in 1900 horses each day produced 1,100 tons of manure, 270,000 litres of urine and 20 carcases (a horse had a working life of 4 to 5 years). One US commentator in 1899 described city streets as "*carpeted with a warm, brown matting of comminuted horse droppings*".[478] All this not only produced a very unpleasant local environment, but there was growing evidence that the mess provided vectors for many bacterial diseases, particularly those associated with respiratory infections. The horse was clearly not a long-term transport solution.

The problem was not new – in 1868 the New York Times had called for "*a cheap mechanical substitute for the horse.*"[479] Thus it is not surprising that in 1898 the Annual Municipal Engineering Conference in Britain issued a statement that "*the introduction and use of efficient motor vehicles should be encouraged (as they) would contribute to the general improvement of the sanitary condition of our streets and towns*".[480]

This concern with clean streets had a strong practical basis, but it was also heightened by the *miasma* theory that was widespread towards the end of the 19th century. *Miasma* was the name used to describe the vapour produced by rotting organic matter and then said to be the cause of most diseases. Roads used by horse-drawn traffic or with road-side sewers or poorly ventilated narrow streets were thus major places where miasma could be encountered. The theory was more widespread in America than in Europe[481] but it certainly had an influence on the City Beautiful movement (Chapter 12b). It was also a strong motivation for cities to invest in public transport systems powered by steam, or cable or electricity (Chapters 12d&e).

CHAPTER 11E

RAILWAYS AT WORK

By the time the London to Bristol railway opened in 1840, railways had become the accepted means of travel within England. Investors were profiting from both the operation of the systems and from the enhanced property values that they created. During the next decade railway capital expenditure grew at an annual rate of 20%. A set of special features that attracted British financiers with their colliery interests was that the new railways bought their coal to the market and in doing so produced steady revenue streams from their haulage prices whilst themselves consuming large quantities of coal.[482]

By 1850 there was an effective national railway system in operation. When the construction phase of the period petered out in 1852, Britain had about 12 000 km of operating track. The resulting rail coverage was extensive but incoherent. The growth of the network continued, but at a slower rate and reached its peak length of about 40 000 km in 1914. There were 21 000 steam-powered locomotives in operation in 1901. Traffic had increased steadily, with growth accelerating after 1870 when fares and freight charges began to drop.[483] Increased demand led to charges rising again and one of the subsequent reactions to the railway monopoly and its higher charges was a resurgent interest in canals, culminating in the construction of the Manchester Ship Canal between 1887 and 1895. Interest had also been sparked by the continued success of the European canal system, which had forced European railways to lower their freight charges.

As network expansion stabilised, the rate of railway investment fell to 5% between 1850 and 1870, and 2% between 1870 and 1890. What had happened is that the high-travel demand routes had been catered for and many newer routes could not deliver in the five profit segments discussed in Chapter 11b – high demand, high value freight, captive market, non-stop operations and land value increases. The investors depended increasingly on nett revenue from day-to-day operations rather than on capital appreciation. However some railways were finding operating profits to be ephemeral and began pleading for government subsidies for the essential services that they were providing.

By the 1860s, as the pace of railway building abated, land owners came to realise that access to a railway enhanced the value of their land. Parliamentary evidence in 1863 indicated that rental values rose by 7% when land was within 8 km of a railway. Thus innovative ways were sought to encourage and then to finance the construction of railways to areas with poor current access.[484]

In a similar context, rather than see all road traffic as competition, railways came to understand the importance of the feeder roads that brought dispersed patronage to and from their stations and freight sidings. For example, Bradshaw's 1863 handbook for railway travellers advises intending travellers on the Great Western services out of London that "*omnibuses leave the City one hour before the departure of each train from Paddington Station which leave the passenger at no loss for a prompt conveyance.*"[485] Thus, although most major roads did fall into disrepair during the railway era, the same cannot be said for many local roads which were well kept and well used. This situation was not restricted to Britain, as Mom[486] demonstrated for Belgium and the Netherlands.

Beasly suggests that British railway operations were never very profitable and that by 1850 at best they were returning 5% on investments.[487] He also concludes that by then the hypothetical steam-powered road carriages to be discussed in Chapter13a would have been more profitable. However, it would seem that investing in the manufacture and possible operation of steam carriages on common roads was far less attractive to investors than was investment in railways with all the associated temptations of rapid capital appreciation, land dealings, property development, monopolies and high-value freight.

Returning to the basic competition between rail and road, typically the horse-drawn inter-town coach services stopped soon after a competitive railway service opened – after all, railway travel was three times as fast and cheaper, indeed sometimes half the cost. The last horse-drawn Royal Mail coach left London in 1846.[488] Thackeray lamented in Vanity Fair in 1847 that the roads had become silent, the coaching inns closed and "*alas, we shall never hear the coach horn sing at midnight, or see the pike-gates fly open anymore.*"[489]

Coach travellers had been predominantly from the upper classes, with cheaper fares for those who sat outside the coach (see, for example, Figs 5 - 10). The lower fares on the railways meant that they also attracted middle class travellers and as the century progressed, travellers on foot on the roads were increasingly from the lower classes and came to be regarded with some suspicion by authorities.

As with the Great Irish Road (Chapter 4c), the case for the first commercial railways had been based on freight haulage and postal services. However, from the time of the Rocket in 1830 (Chapter 10) passenger usage of the new railways far exceeded expectations and it became evident that when a railway opened between two locations, passenger numbers would be at least double the number that had previously travelled the route on foot, on horseback or by horse-drawn coach.[490] Thus passenger travel in the early years of railway was, to everyone's surprise, a major source of revenue although some commentators have since seen it "*as an integral part of the process of industrialisation*".[491] There were four prime sources for these new patrons:

1. transfers across from travel in horse-drawn coaches,
2. business people using personal visits to enhance their commercial operations,
3. families vacationing at more accessible resorts such as Brighton and Bath, and
4. a daily workforce responding to a set of new workplace opportunities.

The fourth category represented a major change-maker, for railways were the first form of mass transportation that could operate on a routine, regular and daily basis. Large industries employing many workers with a range of skills and rewards were now feasible. Workers could live more than 5 km from their workplace.

Some of the early railway services provided passenger seats on the carriage roof, in the manner of the roof-top seating of the earlier mail coaches. More typically, there were separate provisions for first, second and third class passengers (Figs 57 & 58). Typically, first class had numbered and padded seats, second class had covered seats, and third class had uncovered seats in open railway trucks.

The railways had certainly struck a marketing bonanza. Construction could be difficult, but after that consumers were queuing to be served. On land there was no significant competition for the hauling of conventional freight and the carriage of passengers and for both these market segments the growth in the number of consumers was often outstripped by the growth in their appetite for travel. People travelled further to work and to shop and families travelled longer distances for family events and annual holidays. And for the wealthy, there were many new opportunities for conspicuous consumption. Napoléon III in France had his own private train with nine interconnected carriages. European royalty quickly followed suite. In the USA in 1865 George Pullman[492] produced the first of his famous Pullman coaches with their interiors modelled on a first-

Fig. 57 Passenger classes on British railways in 1847. Used in [493] but probably from a contemporary Illustrated London News.

Fig. 58 Lower class travel for passenger on French railways in 1843. Honoré Daumier's view of a train leaving station. From his Impressions et Compressions der Voyage, Le Chemins der Fer, 25 Jul 1843, Le Charivari. From [494].

class hotel and featuring black walnut woodwork and deep-piled Brussels carpets (Fig. 59). They were introduced into Britain in 1874.

The public needed little convincing of the advantages of railway over road. Writing in 1885, Williams commented:[495]

> "*the worst forms of railway travelling for the poorest third-class passenger are incomparably better than the best methods for even the rich of former days. There is now no clambering over dirty wheels – no hurting one's shins on sharp irons – no wedging of one's self amidst piles of luggage on a lofty unsheltered platform, around which numerous legs hung dangling like dozen brace of black and brown grouse; ... no plying of the whip, no foaming mouths, no turgid veins of generous steeds; the giant power that bears us along has bones of brass and iron, and nerves and muscles that cannot tire.*"

Fig. 59 Interior of an 1875+ Pullman "parlor" coach on the British Midland railway From [496].

It has been suggested that the major contribution of the steam railways to the British economy was during the construction phase when they attracted and then distributed large amounts of capital for land acquisition and the construction.[497] However they also greatly aided the coal and iron industries by not only servicing their collieries and factories but also by using much of their output. They lowered travel costs and times for most British enterprises and reduced the need for companies to carry large stocks and operate big warehouses. Farmers had access to a wider selection of markets, although this was often to the disadvantage of local producers. The vagueness of this paragraph largely reflects the absence of a useful review of the economic impact of the new railways on Britain and much data is influenced by the frenzied capital raisings that were associated with their construction.

Whilst the steam-powered road carriage speeds of over 20 km/h (Chapter 9e) were impressive, they were swept away by the Rocket which could haul heavy (and valuable!) loads over long distances at speeds of over 40 km/h. This was also an order of magnitude faster than the speeds that had applied universally to all travel for all of previous human existence. The effect of this widely-experienced quantum leap was to produce extraordinary changes in the way the world operated. Karl Marx in 1857 noted that this investment in transport had caused the annihilation of space by time.[498] The phrase was probably first used in this context in newspaper headlines in April 1838 when the Scottish-built steamship Sirius became the first ship to cross the Atlantic completely under steam arriving in New York just 8 hours before the much larger Great Western (Chapter 9f). Cultural historians and sociologists have observed that the dramatic change in travel created whole new forms of consciousness of space and time which were yet to be noticed within the cities.

At a broader scale, the railway impact on the 19th century was as remarkable as were its local geographic impacts. In the words of Rostow:[499]

> "*Just as the financing and management of the railways set many patterns for large-scale industrialisation on a wider front, so also, it was in the technical experience of building and operating the railways that a good part of the foundations were laid for the march of Western world into (industrial) maturity.*"

I have added (industrial) as I suspect we have yet to reach any semblance of total "*maturity*".

CHAPTER 11F

LOCOMOTIVE TECHNOLOGY

Following the precedent of the *Rocket* (Fig. 53) many operators in the first couple of decades of railway operations relied on four-wheeled steam-powered locomotives (Fig. 60), using up to four locomotives in a row to pull strings of heavy freight wagons. However the Stephensons realised from their initial operating experience that four-wheeled locomotives tended to become unsteady at high speeds. They therefore moved to a multi-wheeled format such as the *Planet* (Fig. 54) with the large driving wheels preceded and followed by pairs of smaller "carrying" wheels. The Planet was the forerunner of a new generation of locomotives and extensions of its design (Fig. 61) were common for the next 50 years.[500] The change is illustrated by the fact that the 1829 Rocket weighed 4½ ton – by 1834 working locomotives were weighing about 17 ton.[501]

Fig. 60 Typical freight locomotive from the 1830s. Made by Bury for the London-Birmingham route.[502] Driving was via the cranked axle of the rear pair of wheels. via Wikipedia and, probably, the Illustrated London News.

Fig. 61 Mid-century locomotive on Britain's Midland Railway. Such six-wheeled locomotives were the common form by the 1840s. From [503].

In the 1840s train speeds had reached about 60 km/h and were limited more by track conditions, inadequate signalling and the need to stop for water every 50 km than by the available power of the locomotive.

Locomotive technology had stabilised by the 1850s. Railway speeds had reached 100 km/h. Indeed, in 1848 Isambard Brunel drove one of his own locomotives, the *Great Britain*, on his broad-gauge Great Western Railway from Paddington to the major branching of lines at Didcot Junction. He took 47.5 minutes, from stationary start to stationary stop. His average speed from start to stop was an astonishing 110 km/h, at least four times as fast as the quickest coach on the same route.[504] In the same year French "*express*" trains were travelling at over 120 km/h.

Locomotive power output was not measured. To do so would have been difficult, given the absence of a drive shaft. Instead, locomotive designers and operators focussed on the tractive effort of the locomotive, measured by the pulling force that the locomotive could generate.

There were no dramatic improvements in locomotive technology in the second half of the century.[505][506] A major commentator has noted that most of the major improvements in locomotive technology had occurred by the 1860s.[507][508] This was primarily because the early decisions on the vertical and horizontal alignment of tracks (Chapter 10b), on the associated clearances from bridges, tunnel walls and track-side objects and on turntable, station and siding lengths meant that there were inherent and inflexible limits to locomotive and train-set size and to the speed at which they could operate. Unfortunately, increasing the locomotive size was the major means for increasing its power output. Another problem was that development was increasing locked into the competitive activities of commercial entities and consequently became secretive and incestuous.

Despite such restrictions, the mid-century steam locomotives and steam coaches (Chapter 13) were to the ordinary citizen, overwhelmingly massive and powerful and beyond any previous human experience of land transport. Such comparisons led many people to be proud of the associated human achievement and Lewis Mumford, a major 20th century commentator, observed that this gigantic size was seen as a symbol of technological efficiency and progress.[509] It also left people feeling individually vulnerable, fragile and dwarfed by the technology.[510]

Although German technology had become a major force in the 1860s, British railways still led the world and for the remainder of the century Britain continued to develop its railway system and to incrementally improve the locomotives that powered it. However, in other areas of transport, it was Germany that provided leadership in innovation and application during the second half of the century. Had British dominance in steam power diverted its attention away from other promising alternatives? Later Chapters return to this theme.

A specific speed-related change in the railway era was in the matter of the displayed times. Given that journeys were now measured in minutes rather than hours, train departures and arrivals were specified to the minute in train timetables, shown on station clocks and checked by train drivers during their journey using on-board timepieces. All these time pieces now had to be synchronised across the country – and telegraphy and radio had yet to be invented. This was a quite new problem for land travellers but it had fortunately been solved in the preceding century as navigators at sea required accurate clocks – chronometers – to determine their longitude and as some standard time had to be established.[511] By the railway era, accurate clocks were readily available and Greenwich Mean Time (GMT) was the British standard. It became the world standard in 1884. Now the time standards of sea captains had to also be applied to land-locked towns. The local town clock had to be set to London-based GMT or citizens would miss their trains!

The main locomotive change in the second half of the century was to take advantage of the stronger steel rails (Chapter 5d) that became available in the 1860s to build yet larger and heavier engines. This required more wheels on the locomotive and from the mid-1860s these multiple locomotive wheels were coupled together (see Fig. 61) to share the wheel-rail adhesion task between all wheels.[512] The concept had been introduced by Stephenson in 1825 (Chapter 9f) but was initially unsuccessful as the connecting links frequently fractured. The coupling problem was later overcome by advances in iron-making and manufacture. The search for better adhesion continued to be largely empirical. Basically, the number of wheels and axles depended in the first place on keeping vertical axle loads low enough for them to be safely carried on rails, track formations and bridges. However horizontal traction increases as the vertical load on the wheels increases and so the driving wheels became large and linked in order to share the piston forces.

The improvement in pressure vessel technology in the 1860s made it practical to produce the more efficient but larger "*compound*" steam engines which used both low and high pressure stages and had first been developed for use on big steamers (Chapter 9f). In the late 1890s German locomotives began to use super-heated steam.

In the 1870s the weight of the vehicles and their payloads continued to increase. Most locomotives had moved from coal to the more efficient coke. Coke also produced a much cleaner smoke. However, there was little competitive pressure to move towards more efficient engines. The world record speed for a steam-powered locomotive was set in 1893 in Batavia, New York State, when the famous 100 ton *Engine 999* travelled at 180 km/h over a measured mile. The limit to steam-powered locomotives at the end of the 19th century was the ability of the firemen to shovel fuel into the furnace as the largest engines were consuming about 2 ton per hour, or 30 kilogram per minute.

Braking had been quite primitive and commonly consisted of wooden blocks which "*brakemen*" would independently screw down onto a small proportion of a train's many wheels. It would take over a minute, or at least a kilometre, to stop a typical train. Thus, locomotive safety was greatly enhanced in 1868 when George Westinghouse[513] invented the air brake which allowed all wheels to be braked by the simple operation of one lever. Stopping times were now less than 20 s, or 300 m.[514]

In the 1880s great rivalry developed between the alternative east and west routes between London and Scotland. This led to an actual railway race between London and Edinburgh in 1888. The Flying Scotsman service won with a speed of about 75 km/h and a time of about 9 h (today the journey takes half that time). A range of locomotives was used, the best was apparently the "*Caledonian 123*" which had a single pair of 2.1 m diameter driving wheels (Fig. 62). Note that the separate supporting frames shown in Figs 54, 60 & 61 as supporting the locomotive engine have now been replaced in British designs by integrated plate or slab framing. However supporting frames remained as a key feature of locomotives operated in the very extensive but more physically demanding US railway system.

When the Firth of Forth Bridge opened in 1895 (Chapter 5d), a further race was held between London and Aberdeen. In this 'contest' the trains were averaging 100 km/h between London and Aberdeen, a record that was to stand for at least another 65 years. It was curious decision as the

British had never seen merit in fast trains due to the infrastructure limits (Chapter 10b) and a management view that 70 km/h "*was fast enough for a gentleman*".[515] Instead, the emphasis in Britain was on meticulous adherence to timetables. Nevertheless, some services did run at 90 km/h and a speed of 130 km/h had been recorded over a 20 km section. The average speed today is about 90 km/h.

Fig. 62 *Caledonian 123* locomotive, built in 1866. It weighed about 40 t. In Glasgow Transport Museum. Photo by Hew Llewelynn, Bristol via Wikipedia Commons.

In 1899 the Grand Central Line between Sheffield and London was opened. It was the first British line designed with an alignment that would accommodate higher speed trains. The service was not popular and the track was later abandoned.

The new internal-combustion engines discussed later in this book (e. g. Chapters 14-16) would steadily replace steam power in the 20th century. The main legacy for the 20th century were "private" transport corridors and all the associated control and service infrastructure developed for the steam-powered railways. Many of these corridors were already able to accommodate a tight schedule of trains travelling at well over 100 km/h and so the gradients were already very flat, the curves very gentle and the stations, marshalling yards and freight facilities were adequate and well-organised. This inheritance would prove to be of particular value for travel within and through densely developed areas.

PART 12

TRANSPORT IN THE CITIES

CHAPTER 12A

RAILWAY TERMINALS

The many magnificent railway stations that were built at the city terminals of railway services throughout the world suggests that the 19th century railway companies had great confidence in the role that their new technology would have in shaping the way the world operated and developed.[516]

As an example, the route from London to Birmingham and the north-west, which was discussed in the preceding Chapter, began at Euston Station, which is 3 km from St Paul's and was then at London's north-western urban edge. The "grand" Doric structure built at Euston Station illustrates the industry's confident attitude (Fig. 63). The decision to build it was made by the directors of the company just 8 months after railway had commenced operation. Demolished in 1962, it was the largest structure of this type ever built. The Curzon St Station built at the Birmingham end of the line in 1842 was of similar overwhelming proportions.

Fig. 63 Euston Station, London terminus of the London-Birmingham railway in 1851. It was built in 1837 and demolished in 1962. Print from Project Gutenberg and Smiles' 1881 Life of George Stephenson.

Many of the new railway stations were, like Euston, built on the urban fringe. Why was this so? Chapter 10b discussed the major disadvantages that the railways suffered due to the severe geometric restraints imposed by locomotive size and the need to use track alignments that were relatively flat and straight. The constraints were particularly felt in urban areas where the railways could not be located within most urban road systems. Thus new urban alignments were uncommon, due to both the relatively high price of urban land and the fact that the private railways lacked the power of compulsory land acquisition. Consequently, the inter-city railways began and ended in cheaper and undeveloped land at the then-current urban fringe.

In the particular case of Euston, the last 2 km of track nevertheless proved very expensive as it also involved a crossing of Regent's Canal.[517] As noted in Chapter 11a, the expenditure was useful as the Canal allowed railway freight to be taken to Thames-side docks. Only a short length at Hampstead Rd – about 200 m from the Station – was easily developed. There was also much devastation of the local housing stock – a process graphically described by Charles Dickens in the opening paragraphs of Chapter 6 of Dombey and Son.[518]

Good further examples of the consequences of this urban-fringe effect are provided by the London ends of many of the new railways. By 1846 there were six railways serving London, all ending at terminals outside the city core and typically in areas where land was still relatively cheap. The terminals were thus often in "less desirable" parts of the city and perhaps explains why the owners of the London-Birmingham railway were quick to invest in a monumental new terminal building at Euston Station. Certainly, the railway users had no misapprehensions and George Bradshaw's popular handbook for railway travellers made the following observation:

> "*The first impressions of a visitor to London are generally of an unfavourable character, particularly if he enter by one of the railway termini, situated in the more thickly populated parts. The dense atmosphere, the squalid appearance of the people, exclude all feelings of pride and admiration from our thoughts.*"[519]

Fig. 64 shows the railways constructed in London in 1845 and 1875 and the railway void in the city core is very apparent. The railways out of London and heading to the north and east initially began at Shoreditch/Bishopsgate – a station a little north of today's Liverpool Street Station and within 2 km of St Paul's. They then passed through the much poorer suburbs of eastern London and into undeveloped land.[520] On the other side of the city, Brunel constructed his broad-gauge Great Western Railway using Paddington as its London terminus. It was in a "better" class – and thus more expensive land – area and therefore some kilometres further out from the City than was Shoreditch. The absence of track in central London (Figs 64 & 65) meant Londoners mainly used the metropolitan components of the railways for excursions and other pleasure trips.

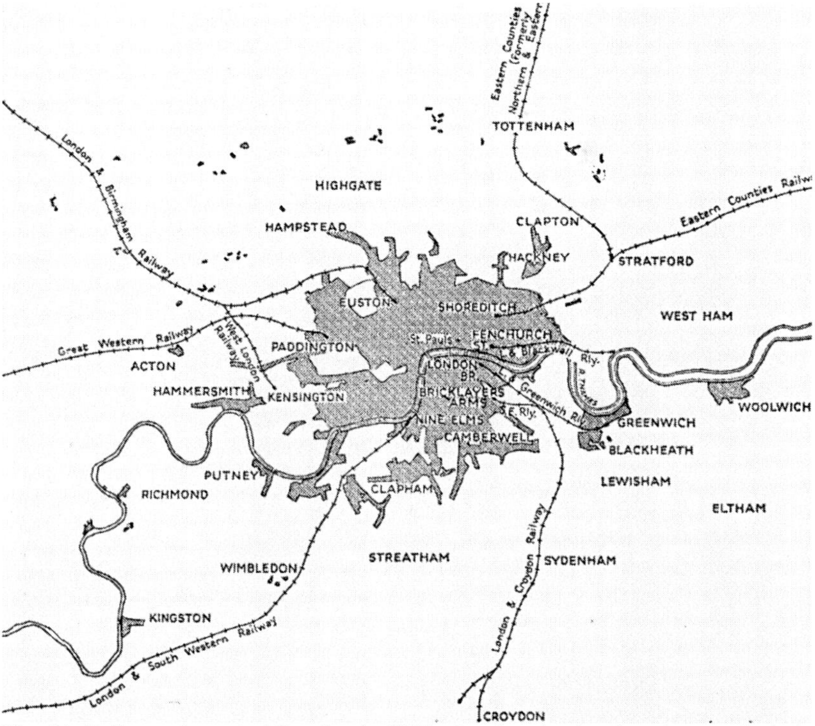

Fig. 64 Railways in central London in 1845, showing the developed areas shaded. From [521].

Although patronage figures can only be estimated, Barker and Robbins[522] argue that in the 1850s work trips by Londoners were still mainly on foot or by horse-drawn bus. They estimate the 1854 daily commuter traffic as 200 000 by foot, 20 000 by bus, 15 000 by steamboat and 6 000 by railway. The population of the whole city was then about 3 million.

A problem for all the railways entering London from the north was a "*horseshoe ridge of high ground that encircled the metropolis on the west, north and east sides*".[523] Prominent features of the ridge include Primrose Hill (Euston Station), Hampstead, Highgate, Hornsey and Stamford. The northern lines out of Kings Cross Station used a gap in the ridge between Hornsey and Stamford Hills.

In 1850 the railway companies were well aware of the commercial value of encouraging city workers to move "*from the dirty suburb to the pure and invigorating atmosphere of the country ... to pleasant and salubrious sites all accessible by railway and within half an hour's ride of the metropolis*."[524] As shown in Figs 65a&b, the railways not only reacted to these market pressures but also created a whole new suburban travel market beyond the old urban fringe. The extension of the northern and eastern lines south from Shoreditch to Liverpool Street Station occurred between 1875 and 1890 and required extensive demolition of existing structures.

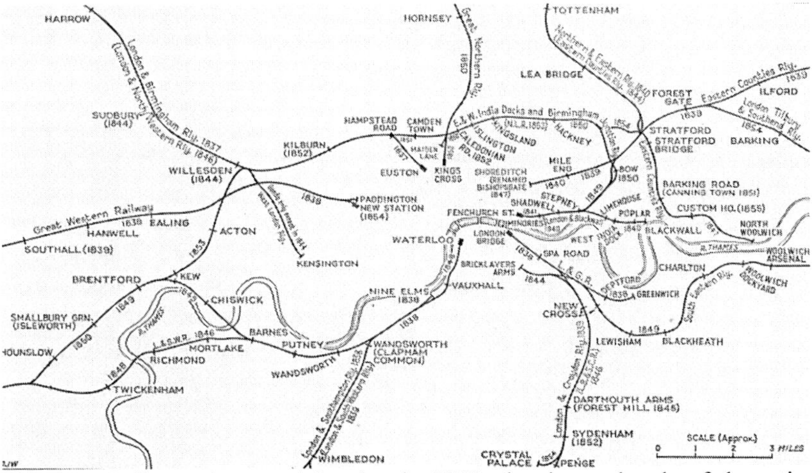

Fig. 65a Railways in central London in 1855, showing a decade of dramatic development. From [525].

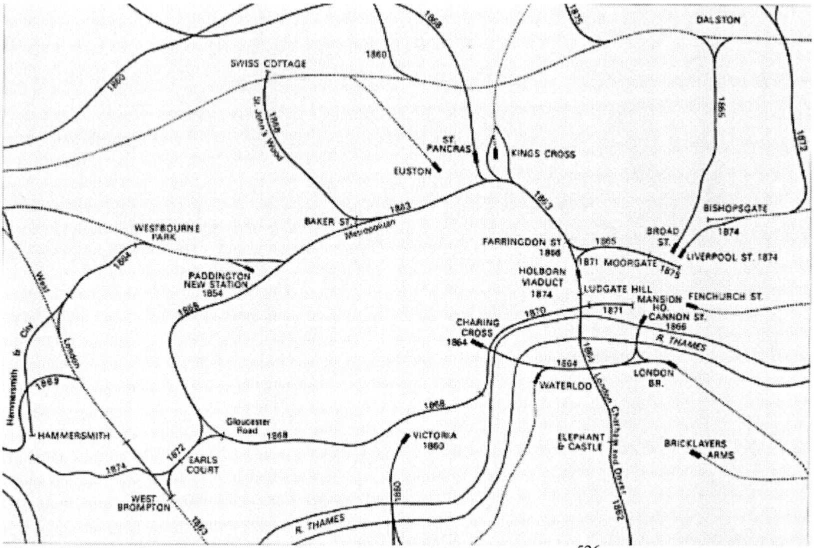

Fig. 65b Railways in central London in 1875. From [526] and used with kind permission of the Institution of Civil Engineers.

The change process continued and by the end of the century the railway companies were very powerful and the demand for their services had continued to increase. New sidings, goods yards and warehouses meant that by 1900 railways were using between 5% and 9% of the land area of Central London, Glasgow, Birmingham, Manchester and Liverpool.[527] However, their relatively high costs of construction and operation meant that the metropolitan railways were far less profitable than the longer rural routes. Indeed, London's horse-drawn bus and tramway companies paid higher dividends than its railways.[528]

There were a number of later schemes to subsidise the fares of travellers from the newly developed suburbs on the then current urban fringe. The stated intent was to help reduce urban congestion, but the logic behind the approach is not apparent as it seems to have been based on a misunderstanding of origins and destinations of the travellers on the congested streets and simply encouraged further outwards expansion.

CHAPTER 12B

NON-RAIL TRAVEL WITHIN THE CITIES

The same geometrical constraints discussed in Chapters 10b & 12a for the countryside meant that the steam railways had no prospect of serving most of the need for travel within the larger cities. However there was a major market demand for intra-urban travel.

The upper class travelled in coaches, both in the countryside and in the cities (Chapter 1f). In London, organised urban transport probably began with the introduction of Hackney carriages to cater for the travel needs of middle class travellers. The two-wheeled horse-drawn carriages were based on French practice and were introduced into London in 1605. The name originally came from the French word *haquence* which referred to the ambling horses used to pull ladies' carriages. Hackneys carried one or two passengers on a route and at a time determined by the passenger. The numbers in operation soon reached plague proportions and so, from 1635, Hackneys were heavily regulated.

It is estimated that in 1800 about eleven hundred Hackney carriages were operating in London. By this time most Hackneys were second-hand coaches bought from upper class owners. In 1805 French Cabriolet coachs (Fig. 12b) were introduced to the London fleet. They were light and elegant and designed for speed and comfort. However, they could carry only two passengers and they were very exposed to the weather.

The four-wheeled Hackney (or "growler") and the Cabriolet were vehicles for the relatively small middle class. It had been commonplace for many people to sleep where they worked.[529] However, the increase in both trade and population during the 19th century and meant that many ordinary workers could no longer live near their place of work. As a result, travel within the cities was quite substantial, and was mainly undertaken on foot. It is appropriate to quote Sekon to describe this period in London. He observed that:[530]

> *"clerks and workers of similar status formed a large part of the big army of pedestrians who twice daily set out on the march (to and from work). The same scene was enacted each morning on all the trunk roads that led*

into the City. The three toll-free bridges (Westminster, Blackfriars and London) were crowded to an extent that is scarcely believable. (At the commencement of the railway era) 100 000 pedestrians a day used London Bridge."

The description has some credibility. Sekon was a pseudonym for George Augustus Noakes (1867 - 1948) and is his name spelt backwards. He was a member of a large London family real estate firm and a railway fanatic who wrote widely under the name Sekon. His descriptions of London transport in the 19th century are based in part on both his own memories and on the extensive discussions he had had with his father.

The *London Bridge* referred to by Sekon was John Rennie's[531] stone arch bridge which opened in 1831. Its traffic way was 15 m wide and, if all the width was devoted to pedestrian traffic in one direction, modern design practice would suggest an hourly capacity of 70 000 pedestrians.[532] However, the bridge was also well used by horse-drawn vehicles (Fig. 66). London Bridge was also the terminus (south of the Thames) for the new railway from Brighton to London and travellers complained that entire 80 km trip took less time than their subsequent 14 km journey across the bridge and on to Paddington.[533]

Fig. 66 Gustav Doré's[534] view of traffic on London Bridge in 1869. The 1831 bridge was replaced in 1968.

East-west travel in inner London was similarly congested as Fig. 67 demonstrates for the Ludgate Hill area (with St Paul's Cathedral in the background). The Figure also shows two horse-drawn buses in the left and centre foreground. These buses were mainly used by workers who had moved to more distant newer suburbs such as Pentonville and Bethnal Green, which were about 2 km north of St Paul's. The relatively large number of pedestrians in Figs 66 & 67 is consistent with the estimate that in the early 1880s the average Londoner made only about 65 public transport trips per year.[535]

Fig. 67 Gustav Doré's view of the London traffic mix at Ludgate Hill in 1869, heading east towards St Paul's. The artist's eye is at Ludgate Circus. The railway viaduct was removed in 1990.[536]

The horse-drawn bus was by then the major form of urban public transport. It was first used when a service aimed at workers was offered in Bordeaux in France in 1812 and quickly spread to other French cities, utilising recent European advances in coach-building (Chapter 1d). The word *bus* arose as a contraction of *omnibus* (latin: *for all*) first used for buses in Nantes in France in 1826.[537] In 1829 horse-drawn buses also began operating along Broadway in New York City.

George Shillibeer,[538] a London carriage-builder, went to Paris to observe the growing coach and bus industry and quickly established his own coach-building business in Paris, adopting and then improving on the recent advances in the styling, suspensions and steering of the Parisian vehicles. After about a decade in Paris, Shillibeer brought his newest bus to London in 1829 (Fig. 68). By carrying multiple passengers, he could offer fares that were a small portion of the existing Hackney taxi fares. However the Hackney industry had been given a monopoly over inner London and his route had to begin at Paddington, outside the area protected by the taxi monopoly.

Fig. 68 George Shillibeer's 1829 horse-drawn bus – the Omnibus. It could carry 29 passengers protected from the weather. The conductor is shown assisting passengers at the rear entry. via Wikimedia Commons.

Horse-drawn buses could only operate at about walking speed and so did not reduce travel times. Their main advantage was that they gave passengers a clean and stress-free ride. The buses proved to be profitable ventures for most operators, although Shillibeer himself left the business and became an undertaker.

By 1830, there were 30 buses operating in London.[539] As discussed in Chapter 9e, there was some minor and intermittent competition from steam-powered buses. However, there was no effective opposition, and some 1000 buses were in service by the time the first length of the Underground railway opened in 1866 (Chapter 14d). The buses needed large wheels to enable them to negotiate rough road surfaces. This restricted the size of the passenger compartment and raised its height, making passenger access difficult (see Figs 68 & 69). The solution to the large wheel problem required pneumatic tyres (Chapter 2) and asphalt paving (Chapter 17c), two inventions that arrived only a decade or so before other factors led to the demise of the horse-drawn bus.

Fig. 69 A typical London horse-drawn bus in Wandsworth in 1894. With permission of the Science Museum London (10650754).

The drivers retained much of the high status that had been gained by the previous generation of coach drivers (Chapter 1e). From about 1850 it became common for some passengers to join the driver and be seated on the roof, or second, level of the buses, as shown in Fig. 69. These outside passengers paid a reduced fare.

The buses were all run by the private sector. The largest company was the Compagnie Générale des Omnibuses de Londre (known as LGO) which had been registered in Paris in 1855, with only three Englishmen amongst its eight directors. By 1856 it owned 600 of the 800 buses then operating in London. The French parent company had been created by Georges-Eugène Haussmann (see below) when he had forcibly merged the ten previously independent Parisian bus companies. The London operations of LGO were centred near Highbury Station, about 4 km north of St Paul's. The control of LGO moved to London in 1859 and from then on the French influence slowly waned.[540]

A major impetus for the urban bus services occurred in 1864 when all of the 81 London toll gates were removed. For travellers this not only lowered the cost of travel, but also eliminated the delays caused by vehicles having to stop frequently to allow tolls to be paid.

By 1900, the combination of electric trolley buses (Chapter 12c) and the metropolitan (largely underground) railway network had replaced many of the horse-drawn bus services. Nevertheless, some 1400 horse-drawn buses were still carrying about 50 000 passengers during each daily peak period. They were being forced out of service, not so much by competition for passengers, but by competition for road space which was being increasingly taken by the growing numbers of cars and trucks.

In 1899 a London bus company placed a double-decker horse-bus body on a chassis powered by internal-combustion and supplied by Daimler's DMG company (Chapter 14b). It was an unconvincing trial, due mainly to problems in maintaining the motor and tyres. More success was had by Richard Stephens[541] who in the same year began operating a 9 seater internal-combustion bus between Portishead and Clevedon in the hills of rural Somerset near Bristol.

From 1834 a major competitor for both the horse-drawn bus and Hackney cab was the new Hansom cab designed by Joseph Hansom, an architect.[542] It was an English development of the Cabriolet (Fig. 12b) which made it more suited to English weather conditions and much safer for passengers, due to its lower centre of mass (Figs 70a&b). The cabs could carry up to three passengers. The low body of the Hansom cab also reduced the risk of passengers being tipped out when the horse fell. Hansom cabs could charge a much lower fare than the horse-drawn buses and in light traffic they could travel much faster, but were initially restrained by legislation. Charles Dickens cynically commented in 1836 "*Why should people be allowed to travel quickly in a cab for 8 p per mile, after Parliament had come to the solemn decision that they should pay 12 p a mile for riding slowly?*"[543] Hansoms nevertheless remained popular for the remainder of the century.

Another response to the increased travel demands within city boundaries was to increase the road space within the cities. Led by London, the largest of all cities of the time, the changes were often dramatic and – despite modern perceptions – long pre-dated the car. New streets which were often brutally cut through the existing urban fabric

Fig. 70a Joseph Hansom's original cab. The wheel diameter was over 2 m. From[544].

Fig. 70b Improved cab in London in 1877. It was still called a Hansom cab, although it was not designed by Joseph Hansom. From[545].

included Farringdon Rd in 1846, Oxford St in 1847, Victoria St in 1851, Queen Victoria St in 1852, the Victoria, Albert and Chelsea Embankments along the Thames in the 1870s, Northumberland Avenue in 1876 and Shaftesbury Avenue and Charing Cross Rd in the 1880s.[546]

Much of the above work was undertaken by the Metropolitan Board of Works, established in 1855 to improve London's infrastructure, centralising power in a single well-funded body. The select committee that led to the establishment of the Board wrote in 1838:[547]

> *There were districts in London through which no great thoroughfares passed, and which were wholly occupied by a dense population of the lowest class of persons who being entirely secluded from observation and influence of better educated neighbours, exhibited a state of moral degradation greatly to be deplored. It was suggested that this lamentable state of affairs would be remedied whenever the great streams of public intercourse could be made to pass through the districts in question. It was also justly concluded that the moral condition of these poorer occupants would necessarily be improved by communication with more respectable inhabitants, and that the introduction at the same time of improved habits and a freer circulation of air would tend materially to extirpate those prevalent diseases which not only ravaged the poorer districts in question, but were also dangerous to the adjacent localities.*

These views continued to be espoused and towards the end of the century many urban renewal approaches came to be embraced in the City Beautiful Movement. It was believed that, although beauty and monumentalism were legitimate ends in themselves, they also would create and enhance moral and civic virtue within the inhabitants of the city.

It is appropriate to also note the development from 1756 of New Rd as a northern bypass of inner London (the *City*) running for about 5 km from Paddington to Islington. Indeed, it was initially considered more of a northern perimeter road creating a barrier between city and country.[548] However new communities such as Pentonville and Somers Town quickly developed outside the "*perimeter*". Today the bypass is inner city route A501. The route was furthered improved in 1826 by McAdam (Chapter 4a) acting for London's new Metropolis Roads Commission.[549] More enhancements occurred when it was rebuilt in the 1860s following the construction of the first stage of the "Metropolitan" underground railway by excavating a trench (the "*cut and cover*" method) beneath much of its road pavement (Chapter 12d). The route was renamed Marylebone, Euston and Pentonville Roads to reflect the new inter-city railway terminals linked by the Metropolitan Line.[550]

During his periods in exile in London between 1836 and 1848, Louis-Napoléon III had witnessed and been impressed by the beginnings of the schemes mentioned above. He saw the Paris he returned to as dirty, cramped and squalid compared with the London he had just left.[551] As Emperor in 1853 he commissioned Georges-Eugène (later Baron) Haussman[552], a career administrator who had studied law and music, to undertake a comprehensive urban renewal of Paris, acting under sweeping powers that he exercised until 1870.

One of the more pragmatic parts of Haussman's brief was to ensure that marching armies could reach any part of Paris and that there was no opportunity for the street blockades of the Revolution (as subsequently immortalised by Victor Hugo in *Les Misérables*). As a consequence, all the streets of Paris are wide and straight without right-angle intersections. Haussman held no brief for the past and to accommodate the major radial roads that had met in congestion on the Île de la Cité in the Seine, he demolished all the medieval buildings that surrounded Notre Dame Cathedral.[553] As with London, these major urban changes occurred before the influence of the car.

CHAPTER 12C

LIGHT RAIL TRAVEL WITHIN THE CITIES

The success of the new steam-powered railways inspired other rail-based developments. In particular, operators saw the advantage of installing railways in otherwise rough and uneven city streets. Low speeds, frequent stops and short vehicles led to the development of much lighter vehicles than those in use on inter-town routes. These new vehicles were called *trams* or *street railways* or *light rail* or *streetcars* in the USA. An advantage of the tram was that it could operate smoothly even when the adjoining road surface was in poor condition. This made it particularly attractive in the burgeoning US cities where the streets were wider but often poorly built and badly maintained. Thus there were horse-drawn trams in Baltimore in 1828, along 4th Avenue in New York in 1832 (Fig. 71) and in New Orleans in 1835.[554] By 1860 New York had 14 horse-drawn tram routes in operation.

Fig. 71 John Stephenson's[555] 1832 New York horse-drawn tram. From[556].

In addition to being able to operate on poor roads, the tram had further advantages. First, its relatively flat vertical track alignment and lower wheel-track friction meant that its operating costs were about half those of a horse-drawn bus.[557] Second, its operating speed was about 30% higher than that of a horse-drawn bus but still sufficiently low to allow it to negotiate the low-radius curves encountered in most city street networks.[558]

Even in the era of the steam-powered locomotive most city trams were still pulled by horses as the locomotives were too large and intrusive for street use. Indeed, in 1840 the citizens of Philadelphia rioted vigorously and violently against the construction of a steam-powered railway into the streets of suburban Kensington.[559] The citizens were unimpressed by promises of enhanced prosperity for greater Philadelphia and far more concerned with pollution and traffic in Kensington. Similar riots had occurred in New York in 1839. A subsequent municipal ban on steam-powered railways was challenged by the railroads but the local Supreme Court upheld the ban, saying that "*steam power would expose the inhabitants to unreasonable perils and might be regarded as a public nuisance*".[560]

The positive aspects of the American experience filtered through to Europe and in 1853 Alphonse Loubat, a Frenchman who had worked in New York City, began operating a horse-drawn tram service in Paris near today's Voie Georges Pompidou (Fig. 72).

Fig. 72 Horse-drawn tram travelling along the River Seine in Paris near today's Voie Georges Pompidou, c1860. From[561].

Horse-drawn trams were introduced into Liverpool in 1859, into Birmingham in 1860 by William Curtis and into London in 1861 by George F Train,[562] an American entrepreneur. The London trams operated along Bayswater Rd west from Marble Arch for a couple of kilometres and ran with flanged wheels on L-shaped rails (Chapter 9c) with the vertical leg of the L raised about 10 mm above the road surface. The arrangement often failed to keep the carriages on the track and also inconvenienced other normal road vehicles. After 6 months, Train was required to remove the entire system. There had been objections from the start, which Train had brushed aside in the spirit of private enterprise, stating "*A man has a perfect right to spend his own money, in his own way, providing he breaks no law, and laws are not yet made for Street Railways*".[563]

Grooved rails became available in 1852 (Chapter 9c) enabling tram rails to be placed flush with the road surface. In the face of growing commuter needs, a dramatic change of heart occurred and in 1871 the British Government approved plans by six private companies to operate about 500 km of tram track in London. The new tramways provided the bus companies with major competition. By 1875 the London trams were carrying about 100 000 passengers a day. The trams and trains were about a quarter of the price of the bus and thus attracted working-class passengers, leaving the buses to the middle classes.[564] By the end of the century there were at least 1500 horse-drawn trams operating in London.[565]

For city services horse-drawn trams remained more suitable than the large, smoky and noisy steam-powered trams. The first major steam trams began operating in Philadelphia in 1860, in Paris and Wantage in 1876 and in Glasgow in 1877. Most used a separate locomotive to pull passenger trailers. They met with considerable public opposition due to their size, noise and smoke. As discussed in the next Chapter, these concerns also applied to heavy rail systems within cities.

Horse-drawn trams were still not widely seen as a means of meeting urban transport demand and many alternative haulage methods were considered. One more acceptable steam-powered alternative was to use cables to haul the vehicles. Recall that the method had been used by some of the first steam-powered railways in the 1830s and 1840s, either completely or on steep grades (Chapter 9d). In 1884 trams climbing Highgate Hill in London were assisted by gripping steam-powered endless cables running in a duct between and below the rails. Such cable-powered systems became popular in many cities. A San Francisco wire-maker, Andrew Hallidie,[566] introduced the method to that city's hilly streets in

1873 and the operation became a model for cable-powered trams throughout the world. By the 1890s over a 1000 km of cable-tram track were operating in American cities. A major inefficiency with the system was that about 80% of the power was needed to merely haul the cables through the ducts, even in the absence of hauled vehicles. Nevertheless, the method was about the cost of the alternative, used in Chicago and New York, of running the trains hauled by steam-powered locomotives on elevated structures built above existing roads.

A number of trials had been conducted in the 1840s, most notably by Brunel, of trains powered by air pressure in pneumatic tubes. The trials had not progressed as adequate seals could not be found. In the 1880s compressed air was used to power the engines of trams in France in a search for engines which were less intrusive in urban areas.

The development of electric power is discussed generally in Chapter 19. In 1879 Werner von Siemens[567] realised that the tram system provided a market opportunity for using his new electricity distribution system. In a trial in Berlin, he initially used an electrified third rail as part of the power network, but this proved unsafe and in 1881 he switched to the overhead wire method now used in most tramways. The first service of this type was in Paris between Place de la Concorde and Palace de l'Industrie. In 1881 Frank Sprague[568] in Richmond, Virginia, USA, introduced the grooved overhead wheel on a stiff pole with carbon brushes to ensure continuous electrical contact between the vehicle and the overhead cable. By 1888 he had perfected the method, which was then rapidly adopted around the world. Britain's first electric tram commenced operation along Brighton Promenade in 1883. Construction and operating costs for electric trams were far lower than the equivalent cable tram service.

By 1903 the US had some 50,000 km of electric tram track in operation. The electric tram was adopted far more quickly than the car and has been described as "*one of the most rapidly accepted innovations in the history of technology*".[569] The reasons for this are clear. Cities were growing rapidly in wealth, population and size. Walking to work was no longer practical, horse-drawn buses were crowded and the stench of horse excrement was becoming intolerable. The new underground railways were expensive to construct and, as they were consequently restricted in their coverage, they satisfied only a small part the market. Thus electric trams captured much of travel market, particularly in the suburban parts of cities. Of course, in the next century their appeal was diminished by competition from motor buses and electric trolley buses and then the car.

Chapter 12a described how steam railways from the 1830s onwards had encouraged the suburbanisation of London. In the 1870s horse-drawn trams "*created the first major wave of middle-class suburbanisation of American cities*"[570] and this continued apace with the advent of the electric trams. America was also quick to develop inter-city electric transport services with about 3 000 km of track in operation in 1900.

CHAPTER 12D

HEAVY RAIL TRAVEL WITHIN THE CITIES

The term "heavy rail" refers to rail systems able to carry their own power supply – typically a heavy steam locomotive. "Light rail" refers to systems such as electric tramways (Chapter 12c) where the in-vehicle motive power comes from a much lighter electric motor and therefore requires much "lighter" infrastructure.

Steam-powered railways were very effective in providing regional, interregional and intercity transport and in meeting the needs of ports and industries. Within cities they were far less effective due to their bulk and noise and clouds of steam and smoke. High city land values also often made the creation of new railway corridors very expensive. One solution was to underground the railways by cut-and-cover construction along street right-of-ways or by the far-more expensive tunnels. In these cases the locomotive smoke created major ventilation problems.

In 1838 the first London suburban steam-powered railway line opened between London Bridge and Greenwich (Fig. 64). Initially the carriages were hauled by ropes and cables powered by stationary steam engines. To avoid level crossings of the street network, about half of the 14 km length was built on a high brick-arch viaduct. It had an adjacent parallel pedestrian path, which was tolled. In its first year the railway carried 500 000 passengers and the path attracted 120 000 toll-paying pedestrians.[571]

The world's first underground railway line was London's Metropolitan Line which was built as a result of increasing congestion on London streets and the lack of any railway service into the heart of London (Chapter 12a & Fig. 65b). The aim was to bring railway passengers to within walking distance of their destination and to provide a real alternative to the horse-drawn buses.[572] The Metropolitan opened in 1863 as a passenger and freight line. Its six kilometre route connected together most of the terminus stations of the regional railway services located to the north of the Thames and then linked them to the inner parts of the City. The route began at Paddington Station which served the west of England, then led to Marylebone Station which then served Manchester, Baker St, Euston Station and the Birmingham service, Kings Cross Station and the northern services and then towards the City at Farringdon St. The major underground length between Paddington and Kings Cross was built by cut-and-cover along Marylebone and Euston Roads and resulted in a quite shallow facility. The section to Farringdon St included about 700 m of tunnel. The tunnels were of a very large diameter, partially to accommodate a broad-gauge rail track (Fig. 73).

Fig. 73 Baker St Station in 1863, part of the first underground railway Guildhall Library, London, image 22510, permission requested.

The locomotives were not supposed to emit steam or smoke in the tunnels, but the air quality was poor and there were many complaints from travellers. Despite improved ventilation and the use of "*smokeless*" locomotives, the problems continued and led to increasing pressures to electrify the system (Chapter 19). The first electric service opened in 1890, running under the Thames from near Monument Station to Stockwell, a distance of about 7 km. Not coincidentally, it was also the first service to run completely underground, operating about 9 m below ground in a tunnel with cast-iron liners.

Subsequently, large steam-powered locomotives did not play a significant role in urban transport.

PART 13

THE STEAM-POWERED ROAD CARRIAGE

CHAPTER 13A

THE MID-LIFE FIASCO

The development of the steam-powered road carriage by Gurney, Hancock and others was discussed in Chapter 9b. The early market success of inter-city steam-powered railways described in Chapter 11e would have reduced the momentum of the steam-powered road carriage industry. Nevertheless, it would have been aware of some major positives that were also occurring. These would have included the success of the new urban bus services (Chapter 12b), the improvements in steam engine technology flowing on from the new steam locomotives, and the need for road-based feeder services for the new railway stations (Chapter 11e). In addition, there would always be trips that could not be serviced by rail. So despite the loss of major market segments, at the beginning of the railway era things were still looking promising for steam-powered road vehicles.

In 1831 Gurney received an order from Charles Dance for a steam-powered carriage to provide a passenger service on the toll road between Cheltenham and Gloucester, a distance of about 15 km. The bus Gurney and his co-investors provided was powered by twin horizontal cylinders placed under the vehicle with the pistons attached directly to a cranked rear axle. It carried six seated and twelve standing passengers and made the journey in a little under an hour.

During the trip the vehicle used 400 L of water and 100 kg of coke. These were relatively large amounts. Indeed, compared to the railways, one of the problems the necessarily smaller road vehicles had was to find enough space to carry sufficient fuel and water and to accommodate the stoker needed to maintain the furnace.

Apart from this supply problem, after Gurney's steam carriage service had been operating for just a few months, a curious amalgam of toll-road trustees, horse-drawn vehicle operators and inn keepers strongly objected to the service. They placed large stones across the road, raised the tolls on the carriage and ordered that no maintenance be done on the road. Consequently, Charles Dance and his fellow investors abandoned the scheme after about 4 months of operation.

The toll roads of the time desperately needed new customers and it appears that their seemingly perverse objections to a potentially major customer had been inflamed by groups of railway investors, and by Gurney's personal support for the local farmers who were petitioning Parliament for lower road tolls. In addition, many toll road trustees had agricultural interests associated with the horse industry. Chapter 1e described how each horse-drawn vehicle needed about 40 hectare of grassed land to support the horses that it used. Thus many agricultural trustees would have had a direct interest in maintaining the operation of horse-drawn transport services, and seen no equivalent benefits flowing from the steam-powered alternatives.

Elsewhere the opposition of the new railways to steam-powered road vehicles was more direct. The Warrington Road Trust controlled a toll road operating in parallel to the new Liverpool-Manchester steam railway. The Railway trustees supplied a clause to the toll road trustees that allowed them to charge relatively high tolls on any steam carriage.[573] These higher tolls – often ten times as high as the existing tolls on horse-drawn coaches[574] – became so extensive that Beasley has argued that in retrospect in the 1840s the railways were a poorer investment than steam carriages, but that the relativity was distorted by the various restrictions, special tolls and legislative restraints placed on steam-powered carriages.[575]

As an example of the legislative restraints, in 1831 a Select Committee of the House of Commons recommended that steam carriages should weigh less than 3 ton, carry less than 15 passengers and travel at under 17 km/h. However, this same committee recommended toll charges for steam-powered vehicles that were more based on the number of passengers or freight tons carried, rather than on whether the vehicle was powered by horses or by steam.[576]

In 1834 the Gurney steam carriages were put into service on a 9 km route between Glasgow and Paisley. After some months of successful operation, a wheel broke on a rough piece of road. The resulting crash caused the boiler to explode and as a result five passengers were killed. Boiler explosions continued to occur and the prospect of further events had been well publicised, as Fig. 74 illustrates. Indeed, as operating pressures rose, boiler explosions remained an ongoing problem throughout the 19th century.[577] By the 1850s, boiler pressures were approaching 1 MPa and many riveted iron boilers were failing in service. British railways averaged at least one major boiler explosion a year with the number peaking at about four a year during the 1860s.[578] Explosions were reduced over time by better means of managing steam pressure and by the use of higher strength steels reducing the number of seams and joints in pressure vessels. Other common problems with steam engines were leaking tubes, broken chains and excessive engine vibrations.

Fig. 74 Robert Seymour's 1829 cartoon "*Going it by steam*" portraying a steam carriage explosion. Seymour was a well-known London artist.

In the face of these strong competitive pressures and community concerns for its safety, rather than present a united front, the steam carriage industry turned on itself. The result was self-destructive and has been described as "*infanticide*".[579] As an example, Walter Hancock was one of Gurney's major competitors in the steam carriage market. In the late 1830s he testified at enquiries that the Gurney steam carriages were too weak and made a noise that scared horses off the road.[580] He was supported by railway engineer George Stephenson (Chapter 9d) who noted the poor stopping distances and smoky exhausts of the carriages, although they did not differ from those of his railway locomotives. Gurney on the other hand counterproductively argued – based solely on the characteristics of vehicles of his own design – that steam-powered vehicles were not suitable for hauling heavy loads at low speeds.[581] Gurney then put steam carriages aside and worked on cures for cholera, lights for lighthouses, and ventilation for the British Houses of Parliament. He was knighted in 1863.

There were many players in the internecine debate, which was wide-ranging and vitriolic.[582] One well-informed commentator from the steam carriage industry later observed[583] "*some of the road locomotionists brought an amount of discredit on their cause by foolishly attempting to rival the railways trains, their extravagancies tending to retard progress of road locomotion.*"

The period has been well reviewed by Mountfield who wrote:[584]

> "*The false dawn of the steam carriage appears as something of a comic interlude in the rolling progress of transport history, an impression heightened by the rolling feud in which the rival inventors engaged. There was relief that the mobile steam engine was confined to railways and banished from the roads. The great iron monsters would not have contributed to the amenities of the city streets or the country roads.*"

Instead of seeing the period as one of "*infanticide*" leading to "*fiasco*", a major English commentator has more charitably remarked that "*during the 60 years following Trevithick's 1803 trial a succession of inventors hammered away ceaselessly at the problem of reducing the size and weight of the steam engine to the manageable proportions required for a vehicle designed to cruise on narrow and tortuous roads*".[585] The new railways had solved the problem by creating their own (rail) roads that were neither narrow nor tortuous.

The railway interests and their associated lobbyists continued to oppose steam road vehicles. For example, in 1859 the Marquis of Stafford and the Earl of Caithness gained some notoriety by travelling through the countryside in steam carriages built by Thomas Rickett,[586] with the essential aid of a driver and a stoker. To stop this process spreading the anti-steam carriage lobby in 1861 succeeded in adding to a Locomotive Bill aimed at making tolls on steam carriages no greater than on horse-drawn carriages, a clause which prevented steam road vehicle from travelling at more than 10 mph (17 km/h) on highways or 5 mph (8 km/h) in cities. In 1865, following debates about public safety, the Bill was further amended to require a man to walk 90 m in front of the steam-powered vehicle and carry a red flag. Thus when steam-powered buses were introduced in Edinburgh in 1869, each 65 seater bus had to be preceded by a man on horseback carrying a red flag.[587]

In the short term, the provisions in the Locomotive Act eliminated any cost-savings that steam-powered road vehicles could deliver to either private or public transport.[588] In the longer term this so-called Red Flag Act was to have major consequences for the development of a British road-vehicle industry. By the time the Act was repealed in 1896 and the speed limit raised to 23 km/h, the age of the steam-powered road vehicle had almost passed and the main beneficiaries were the new internal-combustion vehicles to be discussed in later Chapters.

In introducing the 1896 Bill into the House of Lords, the government spokesman said that the Bill had been drafted in the dark as "*it had no experience of these light locomotives*." There was laughter in the House of Commons when a spokesman there commented that "*it was even possible that these motor cars might become a rival to light railways*." Other speakers feared that the self-powered vehicles would ruin the horse breeding industry. A major commentator noted that the Bill was introduced "*with an air of mild detachment, and the implication that the government had no great desire to take action*".[589]

It is easy to dismiss the initial restrictions in Britain and elsewhere as typical luddite reactions to a new technology, however it must be remembered that the vehicles were often quite large and many times heavier than current vehicles, had brakes that could only slowly stop the vehicle, had inadequate steering, and travelled on roads with alignments that were often unsuitable for even the regulated speeds. In addition, the high centre of mass of the vehicles and the unreliability of their hard-tyred wheels meant that roll-over crashes were always a real possibility. Finally, as noted above, the pressure vessels of the steam engines had some propensity for exploding (Fig. 71). In these contexts, regulatory restrictions were a normal reaction to public experience.

So what if the steam-powered carriage industry had worked in unison rather than in conflict and had received widespread community support? Would steam-powered road vehicles have had the same major economic and social impacts that the railways had in the second half of the 19th century?

The answer to this proposition is "no", for the following reasons. First, by any measure, all the early 19th century steam engines were heavy and had low power-to-mass ratios. When used for motive power in vehicles, the wheel loads would have been very high. Stephenson's *Locomotion* (Fig. 44) had wheel loads of about 2 t which is a little higher than the wheel loads of most current trucks on modern roads.[590] Even if solid rubber tyres or pneumatic tyres had been available, 20th century experience with trucks would suggest that few early nineteenth century pavements would not have withstood many passes of such vehicles without major maintenance. The thinner pavements suitable for coach travel would have proved inadequate and major pavement reconstruction would have been required.

Second, the power output of the engines was relatively low and the railway experience was that they struggled to haul loads up even modest slopes. Thus many existing vertical road alignments would have needed replacement.

Third, earlier Chapters 1b and 8 showed how difficult it is to steer heavy vehicles at low speeds using simple steering mechanisms[591]. At high speeds these problems are exacerbated by the centrifugal forces pushing vehicles outwards, leading to skidding and overturning.[592] Such consequences can be partially managed by:

* good tyre-pavement adhesion,[593] which was not available with then-current unbound pavement surfaces (Chapter 4a),
* road camber,[594] which would however have disadvantaged lower speed vehicles which would be dragged across the road by gravity, and
* the use of very large (c 1 km) horizontal curve radii.[595] Few 19th century road alignments would have been adequate.

Finally, loading and unloading passengers and freight in an effective and timely manner requires facilities separate to the travelled way. These did not exist within the existing road system.

Thus to provide a 19th century road-based service which would have matched that which was provided by railways would have required a similar level of investment in new transport corridors without the appeal of captive paying customers and the associated ease of restricting "free" entry that the railways offered investors during the period of "railway mania".

Perhaps utopia might have been to still create all the new transport corridors, but not to require flanged wheels. Instead the track would be "flanged" or "rutted", allowing steerable vehicles to operate on both the existing road network and on the new corridors. As we saw earlier in the discussion in Chapter 9c on plateways, the flanged tracks provided for non-steerable locomotives had created major difficulties when they became part of the existing road system. The rut-based system used in New York would have been less-troublesome but was discontinued when the operating company found it had an adverse effect on patronage of its own vehicles.[596]

The potential for vehicles to switch between road and a flanged railway system would have been much greater if the pneumatic tyre (Chapter 2) had been invented in the 1820s, as pneumatic rubber tyres are able to develop sufficient interface friction to allow vehicles to be readily turned and braked. A few pneumatic-tyred transport systems were built in the 20th century – for instance in Paris, Ottawa and Adelaide – but used transverse rubber wheels rather than flanges for directional guidance. The other major advantage that would have accrued if rubber tyres had been available at the commencement of the railway era is that the need for very flat grades and large radius horizontal curves resulting from the reliance on friction between iron wheels and iron rails would have been lessened.

CHAPTER 13B

PRACTICAL STEAM-POWERED ROAD VEHICLES

The early successes and failures of steam-powered road vehicles were discussed in Chapter 9b. There had been great public expectations as to what the steam-powered future might look like, as Fig. 75 demonstrates. Nevertheless, Chapter 13a showed that actual progress in developing useful steam-powered road vehicles had been very slow.

Fig. 75 Henry Alken's 1825 view of the future impacts of the steam engine.

Although many and varied political, social and technical reasons have been advanced for the slow penetration of steam-powered vehicles into the provision of road-based transport (e.g. Chapters 13a&b), the reality is that they were not offering any acceptable options for moving people and freight by road. Incremental improvements had not produced the product that the market was subconsciously seeking. On the other hand, it was also becoming increasingly obvious that, despite early enthusiasms, the railway could only satisfy a portion of the needs of the transport market. The world was poised, subconsciously awaiting the next product of the Revolution.

Steam power certainly had satisfied many needs, but it was not without its drawbacks. The steam engines were relatively large and their coal-fired furnaces suggested a practical limit on how small they could be made. Operators were endangered by parts of the engine that became very hot and by moving parts that were difficult to enclose. They also required continuous supplies of relatively large volumes of coal and water. For instance, water consumption was typically 40 litre/km.

As discussed in Chapter 9d, steam engines continued to evolve incrementally by such measures as increasing the area of water pipe exposed to the heat of the furnace and jacketing cylinders to reduce heat loss. A major attraction was that the cost of coke was only about 20% of the cost of maintaining an equivalent team of horses. Thus steam-powered vehicles were becoming more practical and marketable although in mid-century they were still relatively large and ponderous and the practical objections to their use on public roads were real and obvious.

Steering epitomised the problems with providing faster steam-powered vehicles. Recall how Cugnot (Chapter 8) did not discover that his steam-powered vehicle was unsteerable until he took it on its first trial and demolished part of a Parisian defensive wall (Fig. 37). The steering processes shown in Figs 37-39 & 47-50 would also have provided, at the best, a very reluctant vehicle response to a driver's steering input. The lesson was only slowly learnt and when Karl Benz went to take the prototype of the first internal-combustion car onto Mannheim streets one night in 1885, he was unable to turn the vehicle and it crashed into the wall opposite his garage gate (Chapter 14c). It was a lesson well learnt and his 1894 Velo (Fig. 96) introduced the stub-axle forward steering still used in today's cars (Chapter 16a).

The use of a stub axle on each wheel avoids using a single centrally pivoted axle. It had been around since 1714 and had been developed further by Georg Lankensperger[597] in Munich in 1816. It was adopted by many German coach- and carriage-makers but was much more slowly taken up by their English counterparts. It is sometimes called Ackerman steering, after Rudolph Ackerman, a print seller, who was Lankensperger's English agent. The steering wheel on a raked column was first used on the 1898 Panhard-Levassor which is described in Chapter 16b, although the earlier version shown in Fig. 95 does not have a steering wheel.

Understandably the large and unwieldy steam-powered vehicles made little impact on road transport, a circumstance that applied equally to the occasional vehicle powered by internal-combustion (for example, see Lenoir's experience to be described in Chapter 14a). In both cases, engine development had not reached stage where engine power-to-mass ratios resulted in practical road vehicles. The road with all its permutations was a very different beast to conquer than the smooth, well-graded and gentle curves of the iron-tracked permanent way of the railways. By the end of the century, Daimler and Benz had changed this circumstance for internal-combustion (Chapter 16) by dramatically raising the power-to-mass ratios of practical internal-combustion engines. Steam-engine development was not far behind, although hampered by using fuels with lower power-to-volume ratios than petrol (Chapter 15a).

Whilst restrictive regulations (Chapter 13a) and awkwardness had a major impact on the operation of steam-powered vehicles on roads, the use of steam engines for hauling loads and providing mobile sources of power at mining sites, wharves, factories and farms began in about 1850[598] and grew steeply after 1859 when Thomas Aveling[599] introduced a new range of steam-powered devices from his factory in Rochester in Kent, culminating in the Aveling and Porter steam-powered tractor shown in Fig. 76. These were called *traction engines* and were widely used on farms to provide a static source of power for threshing machines and occasionally to pull ploughs. By 1878 there were over 3 000 steam traction engines in operation in Britain. Aveling traction engines were successfully used by German forces in the Franco-Prussian war of 1870-1. Steam-powered machines also greatly improved the effectiveness of hammer-like devices used for the quarrying and crushing of the rock to be used for paving. As noted in Chapter 11a, the first steam-powered shovel had been developed by William Otis in Philadelphia in 1835.

A relevant side benefit of the steam traction engines was that they could be used – as "*steam rollers*" – to provide the high levels of compaction necessary to gain maximum strength and stiffness from macadam pavements and smooth surfaces from the new asphalt mixes (Chapter 17c). Before the steam roller, a typical macadam pavement would be placed in three layers. Each layer would then be subjected to some months of traffic which would compact the layer and allow any ruts to be filled as they formed. The complete process could take a year whereas with a heavy steam-powered roller, it could all be done in a day.

Asphalt paving had been developed in France[600] and Antoine-Rémy Polonceau[601] produced the first steam roller in 1840. Polonceau was also an innovative designer of iron bridges. In England Thomas Aveling began trials of steam-powered rollers in 1865 and the first of his popular line of steam rollers was sold in 1867 (Fig. 77).

Fig. 76 An early 1871 Aveling and Porter steam tractor in the Science Museum, London. The large wheel at the rear provided usable power output via belt drives. Photo via Oxyman & Wikipedia Commons.

The potential for steam-powered road vehicles continued to attract inventors in Europe and North America.[602] However, a negative aspect of the steam traction engines continued to be their very heavy mass, which was applied by their steel-tyred wheels as concentrated local loads, producing high contact pressures on the road pavement. Many road pavements which had been adequate for conventional vehicles were severely damaged by these pressures.

A breakthrough occurred in 1870 when Robert Thompson in Edinburgh invented solid india-rubber tyres to use on a traction engine he had developed to pull a passenger carriage taking part in a bus service trial (Fig. 78). Thompson had previously patented both solid rubber and pneumatic tyres (Chapter 2) in 1846 but had been unable to find rubbers which would enable him to turn his ideas into reality. Pneumatic tyres capable of carrying the weight of steam traction vehicles were not available in the 19th century.

Fig. 77 First production steam roller, made by Aveling Barford for the City of Liverpool in 1867. From [603].

Fig. 78 Thomson's 1870 steam-powered bus From [604].

The growing number of steam-powered vehicles resulted in pressure to ease the Red Flag Act (Chapter 13a) and an amendment in 1878 reduced the distance the man had to walk in advance of the steam engine down to 18 m. The need to carry a red flag was also dropped as it had become the symbol for international communism. Although the restraints were loosened further late in the 1890s, at the end of the century the speed limit on most British roads was still only 20 km/h.

In 1866 the French minister for public works issued a decree restricting the use of steam-powered carriages on roads and limiting their speed to 20 km/h in the countryside and even slower in towns. The decree lasted until 1899. German administrations were even more severe in inhibiting the use of steam-powered carriages on German roads. In 1882 Karl Benz, one of the inventors of the modern car (Chapter 14c), was arrested for breaking a Mannheim ordinance prohibiting the operation of motor vehicles on its local streets. These actions were partly based on the British precedents referred to earlier and also reflected a concern for the impracticality of operating the large vehicles on European roads.

Benz fought against the prohibition and was subsequently given permission to take his vehicle out on test runs, provided he gave prior written notice. In 1893 he was allowed to drive at slow speed without the need for a written notice. In 1896 he succeeded in having the national speed limits of 6 km/h in cities and 12 km/h in the country repealed.

Reviewing developments in US late in the 19th century, McShane remarks:[605]

> "*It is hard to believe that steamers could not have replaced horses for at least some of the variety of tasks that those beasts performed..... they offered enough mechanical reliability, they won races with horses, led circus parades, and demonstrated at fairs. Public suppression was more important than mechanical inefficiency in explaining the failure of steamers*."

Whereas Britain tended to concentrate on the relatively heavy traction engines, the French – perhaps because of a more suitable road system – became increasingly active in producing steam-powered carriages for passengers. Concentrating on passenger vehicles also made it easier to create much lighter vehicles. This gave rise to some very active producers of smaller steam-powered vehicles, particularly Amédée Bollée, Snr,[606] Léon Serpollet[607] and Jules-Albert de Dion[608]. In a major step forward, a number of these manufacturers moved towards using petroleum (Chapter 15a) rather than coal to heat their boilers as the new fuel was far more manageable and volumetrically efficient than coal.

Bollée came from a long family line of coach-builders. He built his first steam carriage, *L'Obéissante* in 1872. It was still very large, weighed five ton and had a separate 15 kW steam engine for each rear wheel, but it could still travel at 40 km/h. The second of his steam-powered vehicles, *La Mancelle* (Fig. 79), was much smaller and pioneered many of the features of the modern car – front-mounted engine, rear-wheel drive with a longitudinal drive-shaft, differential gearing and responsive front-wheel steering. His subsequent series of vehicles continued these advances.

Fig. 79 Bollée's 1878 *La Mancelle* steam-powered carriage, pioneering many features of the modern car. It reached speeds of about 35 km/h. Photo by Pantoine via Wikipedia Commons.

Léon Serpollet began building steam-powered vehicles in Paris in 1877. His vehicles were eminently practical, compared with their predecessors. He greatly improved the design of multi-tube boilers and introduced the capillary "*flash boiler*" which reduced the time required to produce adequate steam pressures and thus made steam-powered vehicles far more practical.[609] Encouraged by Armand Peugeot (Chapter 16b), in

1889 he produced a steam-powered three-wheeled road vehicle and in the same year the pair made a well-publicised trip of about 400 km from Paris to Lyon, with *Serpollet* and his co-driver dressed in formal clothing and wearing bowler hats. Serpollet continued building steam vehicles, although most of his compatriots changed to internal-combustion engines.

In 1883, the firm of de Dion-Bouton was founded by Jules-Albert de Dion, George Bouton[610] and Charles-Armand Trépardoux.[611] De Dion was an aristocrat and a major player in Parisian society, with financial backing from the French branch of the Rothschild family.[612] His two partners were engineers: Trépardoux was an expert on steam engines and Bouton's expertise is discussed further in Chapter 16c. De Dion had seen examples of their work in a toy shop and had then made it financially attractive for them to join him in partnership to build and market a light full-sized steam-powered road vehicle.[613]

De Dion put much emphasis on reducing the mass and increasing the manoeuvrability of his steam-powered vehicles, drawing heavily from technology used in the new bicycles (Chapter 2 and [614]). A steam-powered De Dion-Bouton vehicle (Fig. 80) averaged 24 km/h over a 127 km road competition between Paris and Rouen in 1894 known as the *Le Petit Journal's Competition for Horseless Carriages* (*Concours du 'Petit Journal' Les Voitures sans Chevaux*). It was the quickest trip of all the competitors and was the first time machine-powered vehicles had out-sped bicycles. During this time, de-Dion-Bouton also became active in the development of vehicles powered by the lighter internal-combustion engines (Chapter 14).

Fig. 80 The de Dion-Bouton entry in the 1894 Paris-Rouen road competition in which it recorded the fastest time of all competitors. The form of a powered tractor pulling a passenger coach is similar to Gurney's 1829 steam-powered carriage (Fig. 47). via Wikipedia Commons.

The advantages of the steam engines over the new internal-combustion engines were that they had fewer moving parts, less-demanding machining tolerances and did not need gears to provide a range of power outputs. On the other hand, an underlying problem was that their thermal efficiency at 10% was a very low and left them open to competition from alternative power sources.[615]

Major steam vehicle developments were also occurring in the USA and by the end of the century the Stanley Steamer – built by the Stanley[616] brothers in Maine from 1897 onwards – had captured half of the US road vehicle market. The very neat and manoeuvrable Stanley Steamer shown in Fig. 81 suggests that many practical issues had been resolved. The Steamer used petrol for fuel and had a light tubular chassis. In 1906 a Stanley Steamer broke the world speed record, travelling at 196 km/h.

Fig. 81 The first commercial Stanley steamer, being used by the brothers and as shown in their advertisement in an 1899 issue of Cosmopolitan Magazine.[617]

The steamers best year in the market was 1912, but three practical operating factors weighed heavily against them. First, they were inherently less thermally efficient than internal-combustion. Second, they took about ten minutes to start, as it was necessary to wait until the boiler had "*built up a head of steam*". Third, they then needed relatively frequent stops to replenish the water dissipated as steam. Subsequent improvements in the condensation process would lead to reduced water losses. The "*head of steam*" problem was solved in the 1920s, but by then the market opportunities had been closed by prosperous competitors.

Nevertheless, if the internal-combustion engine had not come on the scene at the end of the 19th century it is evident that the steam-powered vehicles would have satisfied much of the growing market demand for self-powered road vehicles. The only competition would have been from electric power and even a century later batteries able to store appropriate amounts of power still did not exist.

PART 14

THE CHALLENGE TO STEAM POWER

CHAPTER 14A

LENOIR ENTERS THE MARKET

There were clearly strong needs in the transport market which were not being satisfied by either horse-powered or steam-powered systems. There was plentiful power available from the new steam engines but they were very large and intrusive. New players with new solutions were required in order to satisfy the demand for more appropriate engines.

Before entering a detailed discussion of various engine types, it is important to note that the two key engine measures of interest to transport are:

(1) the overall efficiency of the engine and drive system, measured by how much of the energy flow delivered by the fuel can be converted into usable motive power, and
(2) the physical size of the engine per unit of usable power, usually measured by its power-to-mass ratio.

Early engines, particularly steam engines, performed very poorly on both measures.

One of the basic problems with the steam engines discussed in earlier Chapters was that the combustion chamber and the piston chamber required two separate mechanical systems. The combustion chamber would always be relatively bulky as it needed a fuel supply, a fire and a large container to hold the boiling water. For self-propelled vehicle applications there was a clear need for a more compact system and an obvious alternative was "*internal-combustion*" (IC) in which both combustion and induced motion occur in the same chamber.

To the 18th century innovator a possible way to produce the required in-cylinder combustion would be to follow earlier inventors and use gunpowder (Chapter 8) or one of the many combustible gases that were available at the time (Chapter 15). One of the commonest of these was *coal gas* produced by burning coal. For instance, from 1794 in Cornwall, steam-engine inventor William Murdock (Chapter 9b) was producing useful quantities of coal gas, mainly to use for lighting.

Thus, in the 1790s some vaguely reported IC engine trials using coal gas were conducted by Robert Street[618] in Southwark, London. Other trials included the work of John Stevens[619] in New York in the 1790s using alcohol vapour as his fuel.[620] In 1801 Phillippe Lebon d'Humersin[621] in France patented what would have been an inefficient IC engine using coal gas, but died before it could be constructed.

Isaac de Rivaz[622] had been an officer in Napoléon's army before becoming a high-ranking bureaucrat living in Port-Valais at the eastern end of Lake Geneva. There he built gas-fuelled engines between 1804 and 1813, culminating in a self-powered vehicle that was driven in 1813 on a road in nearby Vevey, although it stopped after only 26 m due to a failure in the power transmission system. For fuel he used either coal gas or a gaseous mixture of powdered coal and flammable resins.

Rivaz initially used an electric spark to explode combustible gas in a vertical cylinder, forcing a heavy piston upwards (Fig. 82). As the piston fell back under gravity, it pulled a ratcheted chain attached to a wheel. At the same time, valves allowed spent gases to be forced out and new gases brought into the cylinder, prior to the next spark. Timing of the valve opening and subsequent ignition proved difficult, particularly after Rivaz was forced to change from ineffective electric-spark ignition to ignition using the flame from a lamp. Although the engine was impractical and inefficient, its fuel usage was relatively low and it had some key features in its working cycle that were later to be unknowingly reinvented by others and reappear in their IC engines. It was certainly the world's first IC-powered vehicle.[623]

Apart from the mechanical issue of obtaining useful work from the piston-cum-projectile system used in these initial IC engines, there was also the issue of handling the high pressures produced within cylinder–piston systems made by then-current machining technology. This is the same problem that had plagued the initial steam engine developers such as Watt (Chapter 5c). To avoid high pressures, some followed a method introduced by Huygens in 1673 and used in the early *atmospheric* steam engines (Chapter 8), with the pressure difference between the two faces of the piston head created by condensation of the combustion gases causing a near-vacuum on one piston face.

This was the approach used in the first major British work on IC engines which began in the early 1820s when Samuel Brown[624] in London developed a two-cylinder IC *atmospheric* engine using coal gas. Once ignition had finished, he used condensation of the combustion gases to produce suction. In 1826 Brown employed his engine to power a vehicle which was demonstrated at Blackheath in London. However this simple form of the *atmospheric* process did not deliver sufficient energy to produce a useful power output and so the IC concept was put aside. Brown appears to have taken no account of the lessons previously learnt by Boulton and Watt (Chapter 8) in developing their high pressure, double-acting steam engines.[625]

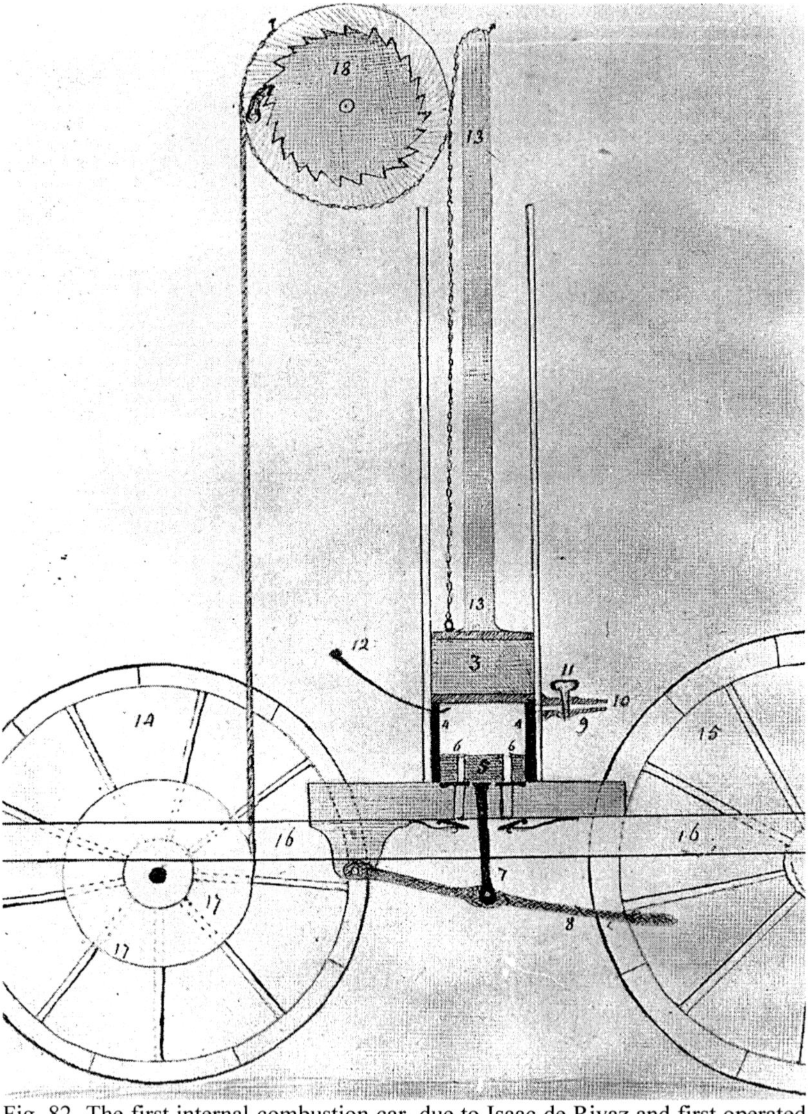

Fig. 82 The first internal-combustion car, due to Isaac de Rivaz and first operated in 1813 in Port-Valais. "Patent" drawing.[626]

A more advanced IC engine was constructed in the 1820s in the USA by Samuel Morey.[627] It had many significant new features, including a carburettor (Chapter 14b) which enabled him to use liquid hydrocarbons as his fuel. However he was unable to commercialise his invention and it "*sank into complete oblivion*".[628] Similarly, Lemuel Wright[629] in London in the 1830s patented a double-acting IC engine based on the Boulton and Watt experience. Despite its potential, there is no record that an engine using the processes described in the patent was ever built.[630] William Barnett,[631] an Englishman, also proposed a double-acting gas engine at about the same time.

As discussed in Chapter 8, atmospheric engines are limited as pressures cannot exceed suction (i. e. atmospheric) pressure. More power could be provided by using the combusting gas to produce a very high positive explosive pressure in the cylinder. As suggested in the Morey and Wright patents, the next steps forward would involve the use of double-acting cylinders in which both processes worked on alternate sides of the piston. For successful operation, the timing and operation of the various ignition and valve processes were of critical importance. Steam engine technology meant that close tolerance piston/cylinder sets were now becoming available and by mid-century double-acting steam engines and the associated valves were well established.

Unaware of preceding developments, two Italians, Eugenio Barsanti[632] and Felice Matteucci,[633] in the 1850s used combustible gas in a double-acting engine. They were the first to use a liquid fuel, naphtha, which at the time was produced as a condensate of natural gas, although in the Middle East it had long been obtained from petroleum and used in warfare. Indeed, "naphtha" was the original Persian word for petroleum. By 1860 the two Italians had built a couple of prototype engines providing about 3 kW of power and are sometimes seen as the first modern IC engines. They also produced a few of opposed-piston engines. No commercial development ensued, possibly because naphtha gas is very flammable.[634]

By mid-century, coal gas was becoming relatively widely available as a fuel for lamps and for cooking (Chapter 15). A major practical engine advance was due Étienne Lenoir[635] in 1859. Lenoir was a Belgian working in Paris. Although he is often narrowly labelled as an "enameller", he had a very wide range of inventions to his credit.[636] This inventive background and a realisation of the potential of coal gas led to him producing the first practical double-acting IC engine

In the Lenoir engine a fuel mixture of coal gas and air was admitted, exploded and exhausted on one side of the piston during one stroke of the piston. An electrical spark exploded the fuel at about the half-stroke position. During the next stroke, the process was repeated on the other side of the piston. This produced an engine crankshaft speed of about 2 Hz, which was close to the limits of contemporary engine-making technology. The effective power output was about 70 W. In many ways, the engine resembled contemporary double-acting steam engines but, as it used the positive pressure of an explosion it was no longer an atmospheric engine.

The engines were inherently inefficient as ignition at half-stroke led to poor combustion of the partially compressed mix and resulted in a jerky piston movement. There was also no provision for cooling the cylinder. Consequently the initial engines were hot and noisy. Lenoir introduced a number of major changes in 1860, including water-cooling of the cylinder. Power outputs of up to 2 kW were claimed but independently measured values were below 1 kW, suggesting overall efficiencies of only 4%.[637] Over 100 engines (Fig. 83) were sold in 1860 – mainly to power lathes, printing presses and water pumps – but users soon found that they were expensive to use and maintain. They also had a seemingly insatiable appetite for fuel and lubricating oil.

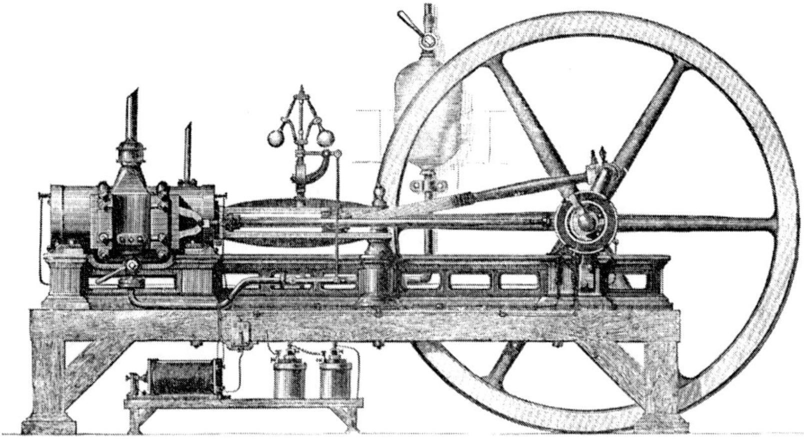

Fig. 83 Lenoir's gas-fuelled IC engine, 1860 version, as in the Musée des arts et métiers, Paris. From [638].

At a public exhibition in Paris in 1860 the engine created great public excitement and speculation about its future prospects. At the time, the widely-read Scientific American magazine reported[639] the somewhat premature Parisian view that "*the steam age is over*". The correspondent saw that the main advantage of the engine was that it used gas rather than coal as a fuel, and that gas was far more "*manageable*" than coal. Not just the media were impressed as Chapters 14b&c will note that perceptive visitors like Benz and Daimler who saw the machine in their travels to Paris were inspired to subsequently take Lenoir's ideas much further.

By 1862 Lenoir was using benzene gas from coal as a fuel (Chapter 15a) and in 1863 he was testing the use of liquid hydrocarbons. He concentrated on stationary (factory) applications and by 1865 had produced about 200 engines. They still operated at about 2 Hz but now had power outputs of between 200 W and a "claimed" 20 kW, although it appears that the maximum achieved was probably closer to 1.4 kW. Others later made the engine under licence. The Lenoir engines were the first IC engines to rival the (external combustion) steam engine as they provided much more power, were smaller and needed fewer staff. Thus they found some favour in small factories.

Lenoir produced three carriages powered by his engines which he called *Hippomobiles* (Fig. 84a). Power was transmitted to the rear axle by a chain drive. The first outing was a 20 km return trip between central Paris and suburban Joinville-le-Pont in 1863, taking about 3 hours (Fig. 84b). However, the power-to-mass ratio of the engines was very low and the vehicles could only travel at 4 km/h. Furthermore, there was no practical way to carry the required quantities of gaseous fuel and water. There is now much scepticism about many aspects of this motorised vehicle.[640]

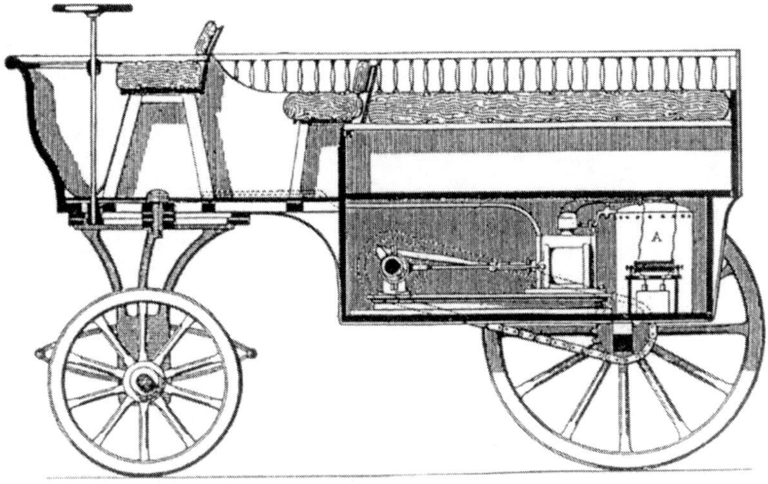

Fig. 84a 1860 drawing of Lenoir's Hippomobile vehicle. From [641].

Fig. 84b Contemporary Parisian view of the first practical trip by an IC-powered vehicle in 1863. From le Monde Illustré?

Despite the promise of Lenoir's work, coal gas was a less than adequate fuel, particularly in terms of available energy per volume, and so progress remained relatively slow. In addition, there were continued incremental improvements in steam engine technology. For such reasons, Lenoir later put further engine development aside. As will become evident, a practical IC engine would not only require better fuel, but also further ingenious invention and improvements in machining technology. Indeed, Hardenburg has observed that by the late 1860s most of the essential elements of the modern IC engine had been envisaged and often designed, but bringing them together in a practical reality would take another quarter century.[642]

The commercial potential of Lenoir's gas-powered engine was not that it would provide a better stationary engine. Steam power was widely and successfully used for this purpose and, although Lenoir had produced competitive gas engines and gas was usually easier to store than coal, the excitement was – as images of Lenoir's Hippomobile (Fig. 84) had shown – about the prospect of future engines with power-to-mass ratios and fuel efficiencies high enough to make self-powered road vehicles a practical proposition.

CHAPTER 14B

SOUTH-WESTERN GERMANY TAKES THE LEAD

The scene now shifts to south-western Germany (Fig. 85). Nikolaus Otto[643] had shown engineering promise as a boy in Holzhausen an der Haide in the 1840s but his widowed mother could not afford to give him a technical education and he became a grocery salesman. Otto was fascinated by a newspaper account of Lenoir's work and began tinkering with the concept in his spare time. In 1860 he used his family inheritance to quit his job as a salesman and concentrate on converting the Lenoir gas-fuelled engine to liquid fuel. In the following year he made a patent application which said that his engine could be used "*for locomotion of vehicles on country roads*".[644] He clearly saw a new market for his "gas" engine.

In 1861 Alphonse-Eugène Beau de Rochas,[645] a French engineer and a friend of Lenoir, patented a process that recognised the explosive value of pre-compressing the gaseous fuel mixture before ignition, but had not constructed any engines using the idea. He also published a theoretical paper on high pressure combustion in 1862. A number of others were also pursuing the same idea.[646]

At the same time and unaware of the de Rochas work, Otto built a model of the Lenoir engine and quickly also discovered the advantage that could be attained over the Lenoir half-stroke process by compressing the gaseous mix in a small combustion chamber between the face of the piston and the end of the cylinder, before exploding the mixture. In this way, the explosive force would occur most usefully at the end of the piston stroke rather than half-way through it. This insight would eventually lead Otto to his invention of the *four-stroke* process discussed below, but his initial approach was to use a single-acting four-cylinder engine in which, at any time, each cylinder would be in one of the four piston positions - or strokes. The four simultaneous positions in the Otto four-cylinder engine are (1) firing the fuel causing a piston to be forced outwards, (2) momentum bringing a piston inwards and exhausting spent fuel, (3) momentum taking a piston outwards and sucking in new fuel and (4) momentum bringing a piston inwards and compressing the new fuel into the combustion chamber. The need for momentum was met by the inertia of the rotating parts, supplemented by the large flywheels used on early IC engines (see Fig. 83 for example).

Fig. 85 Areas in south-western Germany where the competitive development of internal-combustion engines occurred. Most of the region is in the State of Baden-Würtemberg and distant from the industrial centres based in the Ruhr and Silesia. Baden and Württemberg were separate countries between Napoléon splitting of the Holy Roman Empire in 1806 and the new Germany created by Prussian victory in the Franco-Prussian war in 1871.

One of his prototype IC engines exploded in 1862 and his failing finances forced him to turn his commercial attention to making atmospheric engines. In 1864 he met Eugene Langen,[647] a successful businessman with a keen interest in engine development. They created the Deutz Gas Engine firm in Deutz, on the eastern edge of Cologne where Langen's father had a sugar factory. By 1867 the firm was successfully selling an Otto and Langen engine which was only nominally atmospheric as it used explosive combustion to raise a relatively heavy free piston inside a vertical cylinder and on gravity rather than suction to return the piston after firing. The engine was about 4 m high and weighed 2 t and rotated its output crankshaft at about 1 Hz. This was very fast compared with steam engines and indicated that major advances had occurred in manufacturing technology.

Nevertheless the engine was very noisy to operate, required a strong foundation and its power output cost about twice that of the equivalent steam engine. It also needed gearing to reduce the crankshaft speeds to useful levels. However it was at least twice as fuel-efficient as the competitive Lenoir gas engine and received the Gold Prize at the 1867 Paris Exposition. The Deutz engine was soon the market leader for gas engines, selling about 2000 engines per annum. Gradual improvements raised the power output from 350 W to about 2 kW.[648]

Note that a range of values have been quoted for the power output of many of the early IC engines and the figures given in the following text are only indicative as at the time powered output was difficult to measure and reliable data was often uncertain or unavailable.

The smell of invention was still in the air and in 1872 Otto and Langen hired two talented engineers to work at Deutz Gas Engines: Gottlieb Daimler and then Wilhelm Maybach. Gottlieb Daimler (Fig. 86)[649] had left school at 14 with competence in geometry and algebra. He trained and worked as a gunsmith for 4 years whilst studying engineering part-time at an industry training school in Stuttgart. At 19 he was selected to be part of a development program organised by the Württemberg Centre for Trade and Industry. When he returned he spent a year at Stuttgart Polytechnic studying industrial and mechanical engineering. Thus, at the age of 25 he was a skilled and experienced mechanical engineer.

Daimler had become convinced that the ideal engine "*should be ready for operation instantly*"[650] and that there must therefore be a better way than the slow-starting steam engine for producing power for vehicles. Pursuing this dream when he finished at the Polytechnic, Daimler went to Paris in 1860 to study gas-fuelled engines, including meetings with Lenoir. He then worked at two major English equipment manufacturing firms, Crossleys and Whitworths, and inspected British locomotive manufacture at Boydells and Aveling and at the 1862 World's Fair in London, before returning to Germany. Between 1862 and 1871 he worked for various engineering firms in the Karlsruhe – Stuttgart area, and became known for his managerial and his engineering skills.[651]

Fig. 86 Gottlieb Daimler. via Wikipedia Commons.

Daimler and Wilhelm Maybach[652] (Fig. 87) first met in 1865 and were close associates for the rest of their careers. Maybach had trained as a draftsman and worked for a Stuttgart piano maker. The pair moved to Karlsruhe in 1869 where they worked with Karlsruher Maschinenfabrik, a firm making steam locomotives. Karl Benz who was to become Daimler's main rival (Chapter 14c) had been an apprentice at the same firm a few years earlier. When Daimler was hired by Otto and Langen in 1872 as production manager, he successfully recommended that Maybach be recruited to lead the design team. Fifteen years after Otto's initial insight into pre-combustion compression and the associated four-stroke piston process – which would revolutionise the performance of IC engines – Otto in 1876 lodged a patent application for a single-acting, four-stroke engine but concentrated on the first part of his insight – the combustion process.[653] As will be explained below, this was to prove a tactical mistake. Rather than use four cylinders, it was also realised that the four strokes could all occur in the same cylinder with the four basic piston strokes being (1) firing the mix and driving the piston outwards, (2) returning the piston to exhaust the residue through a valve-controlled opening, (3) moving the piston outwards to suck the air-fuel mix though the valve-controlled opening and into the cylinder, and (4) closing the fuel valve and returning the piston to compress the mix. Power is produced on only one of the four strokes. Recall that, at the other extreme, the double-acting steam engine produces power on each stroke.

Fig. 87 Wilhelm Maybach. Photo with kind permission of the Mercedes-Benz Classic Archive (30898).

A major dampener on engine development in France and Germany occurred in the early 1870s as a consequence of the armaments requirements of the Franco-Prussian War of 1870 and the subsequent unification of many German States – particularly those in south-western Germany where much of the engine development was occurring.

All the existing engines in use in the 1870s – including the Otto and Langen Deutz engine – were large and heavy by today's standards. They were certainly too large and heavy to be used to power independent vehicles. The direct key to increasing the power output was to increase the number of piston firings in a given time and consequently increase rotational speed of the crankshaft. By the mid-1870s other firms were selling competitive engines.[654] The Otto, Langen, Daimler and Maybach team at the Deutz Gas Engine factory clearly needed a new product.

As discussed above, the existing Deutz "atmospheric" engine produced an output crankshaft rotation speed of about 2 Hz. The first Otto and Langen prototype using the four-stroke, pre-compression approach was built in early 1876. It had a single 6 L cylinder and was much smaller and a third the mass of its predecessors. It ran at 3 Hz, quadrupled the power output (to about 2 kW), halved the fuel consumption and ran more quietly. Indeed, its power-to-mass ratio brought it close to the level needed to make self-powered vehicles feasible. The engine was thus a remarkable achievement and marked the birth of the practical IC engine. Otto and Langen sold their first commercial version of the engine in October of the same year and were soon offering versions of the engine with power outputs of 1 to 15 kW. They were a major success at the 1878 Paris Exhibition. Within a few years, Otto and Langen engines were being produced with power outputs of over 80 kW.

Otto insisted that the engine be named the *Otto Silent*. However, he became reluctant to increase engine speed further and this limited subsequent improvements in power-to-mass ratios. In 1882 these attitudes and Otto's increasing general conservatism led to the two newer partners in Deutz Gas Engines, Daimler and Maybach, resigning after 10 years with the firm and establishing themselves as a private company called DMG based in Cannstatt (Fig. 85) in a workshop in the garden of Daimler's house. They continued to work on the four-stroke engine despite the prospect that Otto might be granted an exclusive patent.

In 1886 Otto's attempt to patent his four-stroke engine was dismissed. There were two reasons for this. First, as noted above, the patent had concentrated on combustion and given little emphasis to the four-stroke aspect of the engine. Second, opponents of the patent claimed that Christian Reithmann[655] had previously invented the four-stroke cycle.

Reithmann was a Munich watchmaker who had invented major new clock mechanisms. He was also fascinated by engines and combustion at high temperatures and had patented a gas-fuelled engine in 1860. The patent had expired and was soon lost. He may have produced an operating engine in about 1870. When Otto's four-stroke operation became known, Reithmann claimed that his earlier patents and engines had been four-stroke, although no evidence existed to support this. It has been subsequently observed that Reithmann was prone to retrospectively exaggerate his achievements.[656] Further, in his 1861 patent de Rochas had described the four-stroke process with accuracy and clarity, although he had never built a four-stroke engine.[657]

The consequences of the patent decision have been well described as:[658] "*The psychological and practical effect of the invalidation of Otto's patent was similar to that of the expiration of Watt's patent eighty six years before (Chapter 8). It was in fact like the bursting of a dam.*" It was certainly the end of engines which did not compress their gaseous fuel charge before ignition.

More specifically, the patent dismissal left the field open to Otto's ex-partners Daimler and Maybach and to newcomers such as Karl Benz, who is discussed in the next Chapter. It was clear to Daimler and Maybach that the availability of a new high energy-content fuel – petrol from petroleum refining (Chapter 15a) – would further enhance the Otto engine if coupled with a better ignition system and even higher engine speeds, which in turn required better and more heat-resistant lubricants. Coincidentally, such lubricants were also becoming available as products of petroleum refining.

It is useful at this stage to interrupt the story to describe how the gaseous fuel mixes were actually ignited. Atmospheric engines had commonly used a flame to cause ignition, whereas de Rivaz and Lenoir had used electric sparks but these were large power users. The flame approach was impractical at higher engine speeds (above about 1 Hz) and so Maybach in 1885 used a heated tube protruding into the compression chamber. This also had drawbacks; for example the engine could not start until the tube was heated and then had to run at a constant speed. In the same year Benz reintroduced spark ignition. The spark devices still drew too much power and another decade passed until Benz and others introduced *magnetos* which used a rotating shaft (see Chapter 15b) to generate the required igniting spark at precisely the most appropriate time.[659] The invention made higher engine speeds feasible.

In Cannstatt, Daimler and Maybach in 1883-5 built their first engine. Known as the "*grandfather clock*" due to its predominantly vertical and relatively dainty appearance (Fig. 88), in most respects it was the

Fig. 88 Daimler and Maybach's 1885 "*grandfather*" clock engine. It was about a metre high. The piston is within the vertical black cylinder. The lower black disc is the crankcase. The silvery parts at the front of the crankcase provide the rotational power output. Photo with kind permission of the Mercedes-Benz Classic Archive (U2142).

progenitor of today's IC engines. It was a remarkable improvement over any existing product, particularly with respect to its power-to-mass ratio which was at least five times better than any of its predecessors. It had a single vertical 500 mL cylinder and weighed about 50 kg, which was far lighter than any contemporary alternative engine. Its operating speed of over 10 Hz was much higher than had previously been achieved. Modern car engines run at about 80 Hz and the wheels of a car travelling at 70 km/h turn at about 3 Hz. The Grandfather Clock engine had a power output of about 1 kW. The main reason for its relatively high power output for its size was that it introduced fuel into the engine cylinder as a gaseous mixture produced by spraying a fine jet of petrol into the air stream being sucked into the cylinder. The device to do this is called a carburettor. It was invented by Maybach and is still in use in today's vehicles.

In 1886 Daimler fitted one of these engines to a boat. The application was well accepted and from 1886 to 1900 small 200 mL boat engines producing about 400 W were the largest component of DMG's output. In 1888 a *grandfather clock* engine was used to power the propeller of a balloon. Continual design improvements meant that by 1891 the maximum power output had been raised to 2.5 kW. The engine proved very popular and by 1890 about 1900 manufacturing licenses had been issued.

In 1885 Daimler and Maybach put one of their 400 W engines on a wooden-framed bicycle (with outriggers) creating the world's first motor cycle, the "*Reitwagen*" (riding car) (Fig. 89), although a there had been a few trials of steam-engines on cycles.[660] The Reitwagen design is thought[661] to have been based on a Drais walking machine (Figs 15 & 16). In November 1885 Maybach took the vehicle from Cannstatt to Unterturkheim, a distance of about 3 km.[662] The rider was Daimler's son, Paul, who reached a top speed of 12 km/h. [663] Despite the apparent success of the trial, Daimler and Maybach did not pursue the motor cycle idea.

Daimler perhaps perceived the transport future more clearly than did any of the other players that we meet in this story. His most serious rivals were George Stephenson, whose 1821 accurate railway prediction was given in Chapter 9d, and Thomas Telford who thought improved road vehicles were better than steam-powered railways. Daimler saw the need for motive power, he was convinced that steam and external combustion would not meet this need, and he had seen the potential of Lenoir's large IC engine. He was motivated by the need to produce high power-to-mass engines and, once Otto had invented the four-stroke engine, Daimler recognised the potential for firing it with petroleum-based fuels. Moreover, he was not driven by the need to produce motor vehicles for personal travel. His vision was to produce engines not only for cars, but for cycles, trucks, boats, ships, balloons (1888) and aeroplanes.

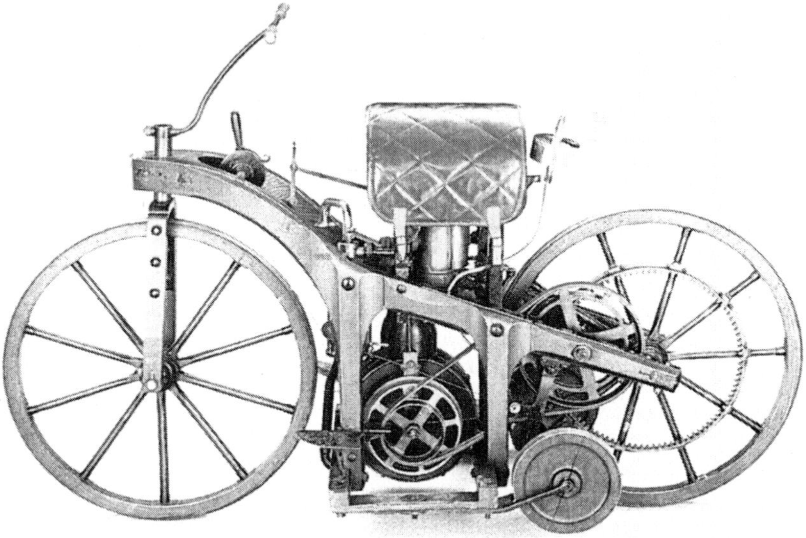

Fig. 89 The first motor cycle. Daimler's 1885 "*Reitwagen*" (Riding Car). Photo with kind permission of the Mercedes-Benz Classic Archive (87F423).

Having made the first motor cycle and motor boat (1886), in 1891 Daimler produced the first internal-combustion truck (Fig. 90), which then dominated the market for the next decade. It is appropriate to note that, although the engine was very advanced, the truck body was a quite basic design derived from horse-drawn vehicles. The engine was attached to the underside of the truck tray and steering was by a pivoted front axle. Their production was followed a year later by the introduction of practical Serpollet steam-powered delivery vans onto Parisian streets.[664]

Whilst aeroplanes might seem a fanciful vision for Daimler, recall from Fig. 75 that powered flight was a matter of speculation in 1825, as steam engines became more common. Daimler died in 1900 but his successors were to produce the engines for the German aeroplanes used in the First World War. However, the first bus powered by internal-combustion was produced by the competing Benz & Cie company in 1895 for a 10 km service between Deutz and Siegen, about 50 km west of Cologne (Fig. 91). It was based on the Benz Viktoria (Chapter 16a) with the larger 3.7 kW engine and could carry 6 passengers. By 1899 Daimler buses were in operation in London.[665]

Fig. 90 The first truck powered by internal-combustion, produced by Daimler in 1891. Note the relatively unsophisticated design of the vehicle body. Photo with kind permission of the Mercedes-Benz Classic Archive (29256).

Fig. 91 The first bus powered by internal-combustion. Produced by Benz in 1895. Photo with kind permission of the Mercedes-Benz Classic Archive (H3000).

CHAPTER 14C

COMPETITION FROM KARL BENZ

Daimler's key rival was Karl Benz[666] who also had a strong steam locomotive background. His father had been a locomotive driver who died in a railway incident when Benz was 2 years old. The young Benz was an outstanding student, attending the local Karlsruhe Polytechnic and then Karlsruhe University, and graduating in mechanical engineering in 1864 (Fig. 92). He then worked for a number of engineering firms, including an "apprenticeship" at Karlsruher Maschinenfabrik which built steam locomotives. Daimler and Maybach had both spent some formative years at the same Karlsruhe factory. Nevertheless, Benz and Daimler never met.

Fig. 92 Karl Benz, co-inventor of the car, in 1869. From Wikipedia Commons.

As a boy, Benz had been fascinated by clock mechanisms. He later said that clocks had taught him "*to feel the marvellous language that gear-wheels talk when they mesh with one another*".[667] As noted in earlier Chapters, clocks and watches had been at the forefront of machine development. The reason for this was that, although their parts were small, they had to be made to precise dimensions and operated at observable and harmless speeds. However, to use cast iron to make geometrically similar but far larger devices able to produce sufficient motive power for transport required tools very different to those used by the watchmaker. A strong early incentive for the development of such tools came from the desire of the military for bigger and better cannons. Clocks and weaponry have certainly played major roles in transport development.

In 1871 Benz established an iron foundry and mechanical workshop in Mannheim. In 1872 he used his wife's wedding dowry to buy out his partner. His wife, Bertha (née Ringer) became an active participant in the work and arguably was the world's first motorist (see Chapter 16a). The 1870s were a time of recession in Germany and the firm struggled for many years and came close to bankruptcy in 1877.

However Benz had also been inspired by Lenoir's engine (Chapter 14a) and, despite his financial problems, he began working on stationary two-stroke gas-fuelled engines. In a two-stroke engine the spent gases are exhausted during the stroke following combustion and the new gas-fuel mixture is brought in during the compression stroke. It requires a more subtle valve system than the four-stroke engine. Benz' prototype operated for the first time on New Year's Eve in 1879, producing about 700 W at 4 Hz. He was still on the verge of bankruptcy when he sold his first engine in 1880. The business then began improving and in 1883 he started a new factory devoted to making his two-stroke engine for industrial customers. The engines were also the first to use spark plugs to ignite the fuel and became the basis of a flourishing business.

In addition, Benz had been a keen cyclist and his "*longstanding dream*"[668] had been a self-powered road vehicle and so, whereas Daimler and Maybach gave priority to the engine over the vehicle, Benz came to place greater emphasis on the characteristics of the vehicle and the four-stroke engine, with its better power-to-mass ratio, became increasingly appropriate for his needs.

Thus in 1883 he sold the two-stroke engine factory and in conjunction with a cycle distributor and a factory owner created a new firm called *Benz & Cie* (in English *Benz & Co*). Whilst waiting for the Otto four-stroke patent issue to be resolved, the firm began making two-strokes with an output of up to 7 kW. These powerful engines were too large to be part of a motorised vehicle. So with the new business running well, in 1884 Benz switched his inventive efforts to the four-stroke engine.[669]

In 1885 Benz produced a new four-stroke engine which ran at about 6 Hz and provided 600 W of power. It had a single 1 L cylinder and was fuelled by commercially-available benzine (Chapter 15a) ignited by electric sparks and cooled by water which had been air-cooled by passing through external tubes and fins (a radiator). He used this smaller engine to power a two-seater tricycle (Fig. 93) in which the engine and chassis were built as a single unit. Like Daimler, Benz had an unsophisticated approach to steering design and had adopted the three-wheel arrangement as it permitted a simple steering mechanism. He surreptitiously trialled the vehicle on the darkened streets of late-night Mannheim. One lesson learnt from these night trials was the importance responsive steering when he crashed into a wall across the street from his garage. Benz henceforth restricted his initial journeys to short trips in his factory yard.

Fig. 93 Karl Benz' 1886 "*Patent Motorwagen*", which was the first public vehicle powered by internal-combustion. Note the large horizontal flywheel at the rear. The chassis drew heavily on then-current bicycle design. See also Fig. 95. Photo with kind permission of the Mercedes-Benz Classic Archive (A90F350).

Immediately after the annulment of Otto's patent in January 1886 Benz was granted a patent for the new vehicle. On 3 July 1886 he took it on its maiden voyage and placed it on public display. This vehicle is discussed further in Chapter 17.

Between 1883 and 1888 Edward Butler[670] at Greenwich in England had developed a similar vehicle. It was also a tricycle and used a double-acting cylinder powered by benzine. He called it the *Petrocycle* but abandoned his efforts because, as he wrote in *The English Mechanic* in 1890, "*The authorities do not countenance its use on the roads, and I have abandoned in consequence any further development of it.*"[671]

The strong push by both Daimler and Benz to produce light engines with high power-to-mass ratios was based on the desire to produce self-powered vehicles, as has been mentioned earlier. After all, in the 1840s Brunel had used 700 kW steam engines to power his first ships (Chapter 9d) and by 1872 Bollée was using a 15 kW steam engine to power road vehicles (Chapter 13b). The inventors' aim was not more and more power but smaller engines requiring less fuel. Primarily this meant faster engines speeds (power is force times speed) and the more efficient use of the fuel's energy. The Benz engine in his *Patent Motorwagen* (Fig. 93) was the first to achieve the above primary objective and it is unreasonable for commentators to describe it as "*a much inferior, slower speed engine*".[672]

In early 1886 Daimler had ordered an "American"-style horse-drawn carriage from a local coach builder as a gift for his wife. When it arrived in August he decided to install one of his new 1 kW *grandfather clock* engines on the carriage. This enabled it to be driven under IC-power in October 1886, making it the first four-wheeled IC vehicle. It was known simply as either the *Motorkutsche* (motor coach) or *Americaine* (Fig. 94). It is clear from the Figure that the vehicle was a carriage designed to be drawn by a horse, to which an IC motor had been attached, as had also been the case with Daimler's first truck (Fig. 90).

Given all the above events, the race between Daimler and Benz to make an IC engine suitable for moving vehicles was probably won by Daimler with his publicly-displayed 1885 motor cycle (Fig. 89). However, the first practical internal-combustion vehicle was Benz' tricycle (Fig. 93) and the first IC car was Daimler's *Motorkutsche* (Fig. 94).

It is interesting to note how the potential of the gas-fuelled engine that Lenoir had invented, manufactured and sold in Paris in the 1850s was perceived and then pursued with great vigour almost exclusively by a number of engineers from a small region in south-western Germany (Fig. 85) – often working in competition rather than collaboration. What led such a perception to be spawned and brought through neighbourhood competition to rapid fruition in such a relatively small part of the world? A German view is that the rulers of the encompassing State of Baden-Württemberg had actively promoted trade in manufactured products and that people of the region were thrifty and diligent and had a compulsion to produce work which was "*as near to perfection as possible*".[673] They had many engineering customers for their engines, but the markets for wider applications of the engines were vastly different. As Chapter 16b will demonstrate, the marketing and commercial application of the new products resulting from their skills occurred, not in Germany, but in France where there was a sophisticated set of consumers motivated by conspicuous consumption rather than by engineering efficiency.

Fig. 94 The first four-wheeled IC-powered vehicle, the 1886 Daimler *Motorkutsche*. Gottlieb Daimler is sitting in the rear and his son Adolf is driving. Note that power is transmitted to the rear wheels via the wheel spokes and is influenced by vertical movement across the rear elliptical leaf springs. Photo with kind permission of the Mercedes-Benz Classic Archive (C23227).

It is also useful to look back at the common sources of inspiration for both Daimler and Benz. Both had worked in the same Karlsruhe factory building steam-powered locomotives. The size and physical presence and brute power of these machines must have been awe-inspiring. But both had also seen how Lenoir, using a new fuel, had produced a far smaller engine which still could produce enough power to move a vehicle. However it had many flaws in its relatively intricate machinery.

Fortunately both Daimler and Benz were also well aware of the intricacies of clock mechanisms and would have known how such devices could be used to provide a system of valves and levers able to control much larger and faster spinning machinery. All that was now needed was some ingenuity, persistence and vision.

PART 15

PORTABLE POWER

CHAPTER 15A

PETROLEUM-BASED FUELS

At this stage, before further examining vehicle development, it is appropriate to pause and consider the new fuels that were now entering the market place. Fuels of course supply the energy that engines use to power transport vehicles. With the major exception of electric motors (Chapter 15b) which can receive their power from distant external fuel sources, transport engines need to have an adjacent supply of fuel. To avoid continual stopping, this fuel must be carried on or in the transport vehicle.

Before the 19th century the only feasible fuels were combustible materials burnt to produce heat which was used to change water into steam, and gunpowder exploded to produce rapid expansion (Chapter 8). Gunpowder was a mix of potassium nitrate (saltpetre), sulfur, and charcoal and had been invented in about 1260. A few other products fall somewhere in the spectrum between slow-burning coal and wood and the almost-instantaneous processes occurring when gunpowder is fired by a spark. Petroleum-based hydrocarbon fuels were a growing part of the spectrum and were gradually becoming the dominant transport fuel, due to their high energy content per unit volume.

Petroleum is a naturally occurring mixture of various hydrocarbons, ranging in form from gaseous, to liquid, to very viscous, to almost solid. It was first found as an oily seepage and sometimes as a more solid bituminous material. From about 3000 BC it had been used in Mesopotamia for a variety of purposes – for instance, as a sealing caulk, as a grease, for covering wounds and as a medicine for digestive disorders. Hydrocarbon oils were also obtained from coal, whales, tree resins (turpentine), olives, sesame seeds, beeswax and alcohol. The vapours from many natural hydrocarbons were highly flammable and were sometimes used for lighting and heating. [674] Other early uses are discussed in Chapter 17c.

As fuels, these hydrocarbon products had attracted little commercial interest in Europe until Aimé Argand[675] invented practical oil-based lamps in 1784. The Argand lamp led to a much increased demand for street and room lighting in the growing and increasingly sophisticated cities of

Europe and North America. Consequently, the demand for hydrocarbon oils rapidly outstripped the existing oil supplies.

It had long been observed that both coal and wood produce flammable gases when heated, well before they burn in a conventional manner. The phenomenon was first formally demonstrated by Johan Becher[676] in 1680 in south-western Germany after visits to collieries in Scotland and Cornwall. Chapter 5b noted that coke was produced by burning coal in a low oxygen environment and Chapter 13b described how coking technology had developed during the 18th century as a more efficient way to use coal and wood for heat production. A bi-product of coke production is "*coal gas*" which is a flammable hydrocarbon gas with the ability to produce a relatively high amount of heat. Tar is also produced as a by-product.

The growing demand for coke from the iron industry and from steam engine users produced relatively large quantities of coal gas and tar. In the 1780s coal gas began to be used for lighting buildings and its increasing and widening availability attracted inventors. In the 19th century coal gas was widely used for street lighting, beginning with a demonstration installation in Pall Mall in London in 1807. As discussed in Chapter 14a, it was also tried as fuel by many of the early inventors and developers of internal-combustion engines.

This was the early days of chemical analysis and one of the findings of these early studies was that coal gas contained so much hydrogen that it became common-place to refer to it as *hydrogen*. Coal gas was also sometimes referred to as *inflammable air*. In the 1820s some inventors used "pure" hydrogen produced by the action of sulphuric acid on metals to power their engines. Although hydrogen as a fuel had many virtues, it was never available in sufficient quantities to be practical.[677] Ammonia was also occasionally used.

In 1823 Eilhrad Mitscherlich[678] in Germany obtained a useful hydrocarbon fuel by refining paraffin – a solid hydrocarbon produced during distillation of coal and petroleum. He named the product "*petrol*". Today it is made by refining petroleum (see below) and that name is still used in many countries. The US equivalent of "*petrol*" is the word "*gasoline*" which is comes from the fact that the original IC engines were gas-fuelled (Chapter 14a). The reliance on a process of chemical extraction meant there had been limited supplies of petrol prior to petroleum refining.

A balsamic resin (or gum) named *benzoin* that had been imported from Indonesia where it is found in the sap of plants of the styrax genus (such as japonicas). It was used to make incense and the name *benzoin* was probably an Arabic term for incense. One more practical use of the resin was to produce benzoic acid which, when heated, provided a hydrocarbon very similar to today's petrol.

The multi-talented Michael Faraday[679] had had a minimal education and at 13 had left school to work as an errand boy. In London in 1825 he invented the process for extracting a similar hydrocarbon (C_6H_6) from coal. Faraday called the chemically-pure product *benzene*. *Benzene* is a component of *benzine* which is a volatile distillated mixture of light-faction hydrocarbons that was (and still is) commonly used a solvent and cleaning fluid. Thus the various *benz*-type names came from the incense and had no link with Karl Benz, although benzene was used as a fuel by Benz (and before that by Lenoir).

The usefulness of fuels such as benzene was quite apparent. They had a relatively high available energy to volume ratio and were easily transported and managed. Consequently the demand for hydrocarbon fuels continued to increase during the 19th century. The demand was finally met by organic petroleum. The viscosity of petroleum is very temperature-dependent and this permits its components to be separated out by temperature-based "*refining*". This began on a commercial scale in California in 1857, using organic petroleum seeping from the ground.

In Titusville in Pennsylvania the locals were obtaining litres of liquid petroleum (oil) by skimming it from the surface of Oil Creek and selling it as *Creek Oil* or *Seneca Oil* to be used as a cure for anything from rheumatism to blindness. In 1858 Edwin Drake[680], an ex-railway man then living in Titusville, was hired to investigate the seepages. He used a team that had previously been drilling for salt and they soon found large quantities of oil 21 m underground. A year later there were over 600 oil drilling companies registered in Pennsylvania.[681] Of course, such drilling was only possible as a consequence of the higher quality iron drill tips available from the new iron works (Chapter 5c).

The various natural oils coming from the drill holes had a range of properties and they needed to be refined and distilled to produce factions appropriate for a particular purpose, from lighting to lubricating. The refining process was based on techniques developed by local whiskey distilleries. Commercial refining to produce kerosene for lighting began in 1865. It was found that kerosene could not only replace the current oils at a lower cost, but would also produce a better light by eliminating flickering. So the search for usable sources of petroleum, particularly in its more manageable liquid form, gained further momentum.

By the 1880s petroleum production and refining was monopolised by John Rockefeller's Standard Oil in the USA and by the Nobel and

Rothschild families creating the forerunner of the Shell Company in Europe. One reason for the Nobel family's interest was that glycerine, a key component in dynamite manufacture, could be made from petroleum. The European market was served by Russian oil.

Initially, the slightly "*heavier*" petrol (or gasoline) was burnt as a waste product. However, it was not long before inventors began to utilise the relatively high energy-to-volume ratios of both kerosene and petrol. The first was Julius Hock[682] in Vienna who in 1870 powered a Lenoir-type engine with petrol. It was only partially successful as the Lenoir engine type needed further major changes. As discussed in Chapters 14b&c, these were initially brought about by Nicolaus Otto, who began working on the task in 1860 at Deutz in Germany and taken to mobile fruition by Gottlieb Daimler and Karl Benz in 1885.

Petrol-powered engines began impacting the IC engine market at the end of the 1880s. By about 1905 they had completely taken over the market. This trend, together with the fact that incandescent lamps – invented in 1878 – were replacing kerosene lamps, meant that petrol became an increasingly important product of petroleum refining.

CHAPTER 15B

INVISIBLE POWER

The solutions to providing power so far discussed have all involved burning fuel in the vehicle being moved. In this process a lot of the energy produced and the space being moved is consumed by the fuel needed for the journey. What if the fuel could be burnt in a fixed location and only power itself transported to or by the vehicle? The answer to this question takes the 19th century changes through another quantum jump, for all the inventions discussed to date fit into normal understandings and intuitions (except, of course, for the matters of vehicle speed and engine component speed). They were clever and innovative but followed logically from previous understandings. No new truths had to be unveiled. Perhaps the closest to newness was creating a vacuum by rapidly cooling steam, but this could have been deduced by observant tinkering over a kitchen stove.

Electricity was quite a different matter. Before the 19th century it was known mainly via natural lightning displays of static electricity and serious and productive research into static electricity had occurred in the 18th century. At a human scale, electric charges were produced by simple techniques such as rubbing objects together although how and why this happened was not understood. Taking a different tack, in 1800 Alessandro Volta[683] at Pavia in Italy invented a device (a *battery*) which produced electric charges by the electro-chemical reaction between layers of zinc and copper or silver. The layers were stacked vertically and the (dry) device became known as a *voltaic pile*. It was the first device to produce a continuous flow of electricity.

The major intellectual jump occurred in 1820 when Hans Christian Ǿrsted[684] in Copenhagen observed that a magnetic needle (as in a compass) was deflected by electricity in a wire from a battery, without any apparent physical contact between the wire and the needle. This led over a couple of decades to the linking of electricity and magnetism and the visualisation of an invisible *magnetic field* surrounding each pole of a magnet. The field was an almost mystical concept and required a great leap of faith, particularly as there was some dissonance between this concept and the understandings that had developed around Newtonian gravity.[685]

Moving a delicately suspended compass needle might seem far removed from producing usable motive power but Michael Faraday, working in London, immediately took Ǿrsted's ideas forward and in the early 1820s invented electromagnetic rotary devices which could interchange electric power and rotary power. He also documented the vital process of electromagnetic induction whereby power is induced in a wire by movement of a magnet. However the concepts involved remained difficult and Faraday himself only introduced and utilised the descriptive power of the term *magnetic field* in 1845.[686]

In 1825 William Sturgeon[687] in Surrey took the concept a stage further by wrapping a wire carrying a current around a piece of iron to create a magnet – indeed, an *electromagnet.* By 1832 Faraday and others had produced laboratory devices (generators, dynamos) capable of generating power from rotary motion and, in reverse, motors capable of generating motion from electric power. The induction process was quite alien to human intuition and was not given a firm scientific basis until James Clerk Maxwell's[688] work in the 1860s. Fifty years were to pass between Faraday's experiments and the use of the strange new concept for major power production. Of course, the intellectual understanding of the electromagnetic field by Faraday and Maxwell would have impacts on civilisation which extended far beyond transport.

Faraday was not alone in his interest in electric motors and there were many inventors attracted to the concept in the 1830s and 1840s, not only in Britain but throughout Europe and in the USA. Initially the devices mechanically resembled the beam-type steam engines (Chapter 8 and Fig. 36) but these were gradually superseded by motors with a direct rotational output.

A curious and useful feature of these "motor" arrangements was that applying electric power to the device caused usable relative movement of the components – it was then an electric motor. However, by 1833 it was realised that externally causing relative movement of the components produced electric power – the same device was then an electric generator.

Moritz Hermann Jacobi[689] was one of the more successful earlier electric motor developers. Whereas many early engines produced only a few Watts of power, in 1838 Jacobi was using a 300 W motor to operate an 8 m long boat on River Nerva in St Petersburg. In 1836 Robert Davidson[690], a chemist in Aberdeen, had used non-rechargeable batteries to power electromagnets which pulled bars of iron attached to a drum. In 1839 the rotation of the drum was used to propel a vehicle which was probably the first practical electrically-powered car. In 1842 he used a 700 W battery-powered electric motor to pull a vehicle at walking speed along the Glasgow to Edinburgh railway line. The main limit on all these motors was the low power output and the short operating life of the batteries available at the time.

As a consequence of the improving motor technology, battery development proceeded at some pace. In 1836 John Daniell[691] introduced the wet, or Daniell, Cell in which a zinc strip was placed in a bath of sulphuric acid held in a copper pot. It only produced about a volt but was far more reliable than the earlier dry cells. The batteries that followed were relatively large and bulky as they used chemical processes to store electricity. They were also not rechargeable, highlighting the then current problem of storing electric power once it had been generated.

The new railways provided the incentive for the next development stage, but not in a quest to provide motive power. The electric telegraph was invented by Paul Schilling[692] in Cannstatt (Fig. 85) in 1832 and used on a German railway in 1835. The first practical use[693] was from Paddington to West Drayton on the Great Western Railway in 1837 using a method developed by Charles Wheatstone[694] and William Cooke.[695] Indeed, it was arguably the first practical use of electricity. The telegraph also had some impact on the need for physical forms of postal transport. However, its major initial use was to manage railway operations. The efficiencies in scheduling were very obvious and telegraph lines were quickly installed along most railway tracks.

The telegraphs needed reliable sources of electric power to carry the signals. One of the first practical applications of electric generators was by Werner von Siemens in 1856. He developed small generators which allowed the telegraph operators on the Bavarian Railways to generate the required burst of power by rotating a handle at the end of the generator. He realised the importance of the geometry of the electromagnets – or armatures – with particular emphasis on the way in which the wire was wrapped around the iron.

It was realised that permanent magnets were not an essential part of the devices as they could be replaced by further electromagnets which could move relative to other electromagnets. The development and manufacture of practical and more powerful electric generators and motors was a much slower process as it depended on advances in manufacturing technology and did not come on the market until the 1870s after von Siemens, Charles Wheatstone and Samuel Varley[696] in London in 1866-7 produced dynamos using electro-magnetic fields rather than permanent magnets. By 1870 generators were being marketed which provided a powerful direct current output at a steady voltage.

It was now feasible to distribute practical levels of electric power by wire. The potency of this idea was first seen when it was applied to electric tramways, as has been discussed in Chapter 12d. In 1887 Sebastian de Ferranti[697] in London demonstrated the far greater efficiency of distributing power from generators to users by employing alternating rather than direct current although Edison argued strongly for direct current transmission.[698] The other great change was Charles Parsons' 1884 redesign of the steam turbine discussed in Chapter 8. This made it possible to generate large amounts of electric power far more cheaply than had previously been possible.

The development of rechargeable batteries by Raymond Planté[699] in 1859 made it theoretically feasible to use batteries in routine transport operations and Nicolas Raffard[700] made a battery-powered electric car in Paris in 1873. Further developments occurred in France but the first practical electric car was built by Thomas Parker[701] in London in 1884. He had previously electrified the London underground railway system and tramways in Liverpool and Birmingham.

Batteries were relatively heavy and bulky and not a practical transport solution for independent vehicles. Nevertheless, in 1897 about 12 electric taxis were operating in New York and some trial battery-powered buses and taxis were tested in London. In 1898 the Pope bicycle-making company in Connecticut began the commercial production of electric cars. There was a common view that electric cars would soon supplant vehicles powered by internal-combustion due to the various external appurtenances needed to make IC vehicles operate efficiently. However, the contemporary storage batteries made electric vehicles too heavy to be practical for every-day use. The transport role of electric vehicles in the latter portion of the 19th century is discussed in the next chapter.

In summary, the development of electric motors and generators followed a different pattern to the development of internal-combustion engines discussed in earlier chapters. The electric devices followed rather than preceded science and were, of themselves, relatively simple devices. They were also relatively efficient and with good power-to-mass ratios. Their drawback in a transport context was that there was no easy way to provide the energy input to the motors of free-ranging vehicles.

Two very important related inventions in the last quarter of the century were the creation of the telephone by Alexander Graham Bell[702] in 1876 and sound recording devices by Thomas Edison[703] in the following year. In 1895 Guglielmo Marconi[704] in Bologna, Italy, invented the radio and over the next five years conducted a number of increasingly impressive demonstrations. These inventions developed to a useful level in the new century. They would dramatically change the way in which the 20th century world communicated and thus significantly modify its transport needs.

PART 16

THE MARKET FOR THE NEW CARS

CHAPTER 16A

BENZ AND DAIMLER BECOME SERIOUS CAR-MAKERS

In January 1886 Benz received a patent for his tricycle, which came to be called the *Patent Motorwagen*. The vehicle was described in Chapter 14c and shown in Fig. 93. Its 700 W of power was delivered to the driving wheels via a sprocket and chain connected to "*differential*" gearing on the drive axle. The power was sufficient allow the vehicle to travel at 15 km/h but, in the absence of gears it had difficulty climbing hills. The engine was capable of running on benzene, naphtha and the newly developed petrol (Chapter 15a).

Differential gearing allows the wheels on each end of a drive shaft to revolve at different rates as a vehicle is turning. It is essential if the vehicle is to follow the trajectory needed to negotiate a conventional street system. The process had long been used in clock mechanisms. It required careful machining and so it is not surprising that it was first applied to steam powered vehicles in 1833 in Manchester by Richard Roberts[705] who had earlier made the first practical machine tools for gear cutting. In 1876 James Starley (Chapter 2) developed a chain-drive differential for his tricycles and this approach was used by Benz.

Benz was aware that his vehicle was underpowered and worked to add gearing and increase the power output of the engine. In 1887 he produced Model 3 which had a 2 kW engine and wooden rather than wire wheels. It could travel at up to 20 km/h and manage slopes of 1 in 10. Benz presented the vehicle at the 1887 Paris Exhibition and in early 1888 he was visited by French engineer Émile Roger[706] who was already an active sales agent for the Benz stationary engines. Roger bought one of the vehicles together with French sales and manufacturing rights for subsequent vehicles. It was the first commercial sale of an IC vehicle. About twenty-five Model 3 vehicles were sold, mainly named *Roger-Benz* or *Rogers*, to avoid Franco-German animosity.[707] Early in 1888 Benz exhibited his vehicle at an Exhibition in Munich, but sales continued to be too low to make the venture profitable.

In August 1888 the world's first ordinary car trip was made by Bertha Benz, Karl's wife, when, accompanied by her two young sons, she drove the car about 100 km from Mannheim to visit relatives in Pforzheim, buying fuel en route from local pharmacies. The success of the *Fernfaht* (long drive) was a major morale booster for Benz and his supporters and the vehicle was exhibited at the 1889 Paris Exhibition. However, the attendees at the Exhibition were far more interested in the new Eiffel Tower than in than in the strange vehicles separately exhibited by Benz and Daimler. Similarly, the external partners in Benz & Cie were unimpressed with Benz' increasing emphasis on vehicles rather than stationary engines and withdrew their financial support.

In 1890 two new partners joined Benz & Cie and vehicle production came an increasingly large part of the business. For most of the 1890s Benz & Cie was the world's largest manufacturer of IC vehicles and in 1900 it was producing 600 vehicles per year. However Roger in particular was unimpressed with the commercial potential of single-seat three-wheelers and passed this view on to Benz.

Thus, in 1893 Benz introduced the *Viktoria Ideal*. With more available power, the design could be based on the heavier horse-drawn carriages rather than on the bicycle technology used in earlier Benz vehicles. *Victoria* was the name of a common late 19th century horse-drawn coach (Fig. 95a). The design similarities between the two are apparent from Figs 95a&b. The four-wheeled Benz vehicle could seat three passengers (orse-drawn)including the driver. It had stub axles and was steered by a vertical steering wheel. It was much more powerful than the original *Patent*, as its single 1.7 L cylinder produced about 2.2 kW. The *Ideal* was later replaced by the *Viktoria Comfortable* with a 2.9 L engine producing 3.7 kW at 10 Hz. Roger's continuing advice again proved correct and 45 *Viktorias* were sold in 1893. They could be demanding cars to drive as one hand was needed for steering, one for the throttle and a third for the gear change.

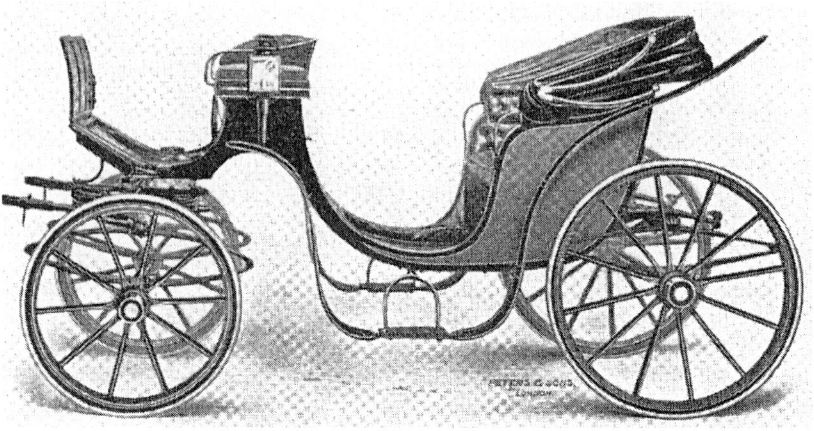

Fig. 95a A c1880 horse-drawn *Victoria* coach. It was derived from the buggy (Fig. 13) but had much smaller wheels. From [708].

Fig. 95b The 1893 Benz *Viktoria*. Note the two front wheels with stub-axle steering (Chapter 13c) and a small vertical steering wheel. Photo with kind permission of the Mercedes-Benz Classic Archive (2003DIG193-33).

In 1894 Benz began making the first production-oriented motor-vehicle, the *Velo* (Fig. 96). This was much smaller and less powerful (1.0 L producing 1.1 kW at 7.5 Hz) than the *Viktoria* but had three forward gears. It was otherwise very similar to the *Viktoria Ideal*. It sold for about half the price of the Viktoria which made it relatively popular and, at the end of the century, led to Benz & Cie continuing to be the world's largest car manufacturer. By 1902 about 1200 Velos had been sold.

A larger version of the Velo (2 cylinders of 1.5 L providing 5.5 kW) was produced and in 1895 buses were also being manufactured (Fig. 88). The available engine power was now about 7 kW and this permitted vehicles to travel at up to 30 km/h. It has been said that this was "*about the limit that the current Benz transmission and chassis details would stand*".[709]

Consistent with his conservative view of the market, Benz recommended against using pneumatic tyres because, as the 1900 Benz & Cie catalogue stated, "*they skid badly and are liable to puncture.*" Introduced for bicycles (Chapter 2), their first use on cars had been when the Michelin brothers, Édouard[710] and André,[711] used them on their entry in the 1895 Paris-Bordeaux road race to be discussed in Chapter 16c. They had improved on Dunlop's invention by adding wire-beaded edges to the tyres. A year later pneumatic tyres were being used on de Dion-Bouton and Bollée production cars (Chapter 16c).

Fig. 96 The 1894 Benz *Velo*, the world's first production car, driven by his daughter Clara. Photo with kind permission of the Mercedes-Benz Classic Archive (1989M163).

Benz had completely misjudged the usefulness of the new tyres which were soon to have a major impact on the effectiveness of motor vehicles. They greatly improved rideability as a consequence of their more than adequate flexibility and had the added advantages that:

(a) the contact pressure between the tyre and the surface it was running on could not exceed the inflation pressure of the tyre, thus reducing damaging high stresses produced by the stiff solid tyres,
(b) the tyre could form itself around small obstacles thus reducing any vertical movements transferred to the vehicle, and
(c) the tyre surface provided good frictional contact with all common pavements, thus catering for the turning, braking and accelerating demands of a speeding vehicle.

Despite the success of Benz & Cie, in the 1890s the centre of motor vehicle initiatives moved from Germany to France. The move related to technical, financial and personal matters which will be explored below. Indeed, by the end of the century about six French firms, such as Hurtu, were manufacturing Benz-type cars. Even Benz' motors were being built in Ardennes in France – about 300 km west of Mannheim – and the bodies in Paris. During this period Benz had become resistant to major change and, for the traveller, the solid-tyred 1899 *Velo* – although now powered with a 7 kW engine – still had the open and exposed characteristics of his 1886 *Patent*. The fact that it had also gained a deserved reputation for reliability and ease of maintenance was insufficient.

During this period Daimler and Maybach were more interested in engines than in the vehicles in which they were installed. Maybach in particular made many innovative engine improvements – such as fuel injection and crankcase compression.[712] In 1889 they modified the 1886 *Motorkutsche* (Fig. 94), using bicycle rather than horse-drawn coach frames, substituting wire-spoked wheels and installing one of their new engines, which produced about 1 kW at 10 Hz. It also used a four-speed toothed-gear transmission. Now known as the *Daimler Wire-Wheel* car, it was exhibited at the 1889 Paris Exhibition. It symbolised the change from horse-drawn to IC power, but gave only passing acknowledgement to customer needs, a thought that must also have been in the mind of the two riders of a *Wire-Wheel* shown in Fig. 97.

The technical success of Daimler and Maybach had not led to larger market sales for their stationary engines, so Daimler had to seek outside capital to expand production. Two of the financiers he aligned with had made their money in munitions and the third was a banker.[713] The refinancing process culminated in 1890 with the establishment of Daimler-Motoren-Gesellschaft (DMG) as a public company

However, things were about to change. In 1891-2 DMG began producing a new engine appropriately called the *Phoenix* but still based on Daimler and Maybach's groundwork. The initial version had two parallel vertical cylinders and a power output of about 2.5 kW at 12 Hz. A second version had the cylinders in a 15^0 V configuration. This "*V-twin*" engine was based on Daimler's last major patent, which he had obtained in 1888. Daimler and Maybach became increasingly unhappy with DMG corporate decisions. Both left the company: Maybach in 1891 and Daimler in 1893. In 1891 Daimler had separately and wisely sold his patents to Panhard-Levassor for France, to Frederick Simms for the British Empire (see Chapter 16b) and to William Steinway of piano-making fame for the USA. He also retained the commercial use of his name. After Daimler left, the company became increasingly less profitable.

Fig. 97 The Daimler *Wire-Wheel* car at the Paris Exhibition in 1889. The two occupants are Renée Panhard and Émile Levassor who were soon to take the next major step in the development of the car, as discussed in the next Chapter. An alternative name for the vehicle was the steel-wheeled car (*Stahlradwagen*). Like the Benz *Patent Motorwagen* (Fig. 93), its chassis also drew heavily on bicycle design. Photo with kind permission of the Mercedes-Benz Classic Archive (47292).

As with Benz, Daimler had a conservative view of the new vehicles. He did not believe that drivers would be able to handle faster cars powered by bigger engines. However, Maybach thought otherwise and continued his innovation. In 1890 DMG began producing the world's first four-cylinder engine, with a much enhanced power-to-mass ratio. Nevertheless, the new part-owners of DMG placed even more emphasis on stationary engines rather than on cars.

CHAPTER 16B

THE FRENCH MOVE INTO THE MARKET

René Panhard[714] was a French engineer who had graduated from the prestigious École Centrale des Arts et Manufactures in Paris. He began his career in a firm which made wood-working machines and was owned by Jean-Louis Périn. Panhard became a partner in the firm in 1867 and helped turn it into one of the largest engineering facilities in Paris. In 1872 his École classmate, Émile Levassor,[715] also became a partner.

A key player in the firm's development was Edouard Sarazin, a Parisian lawyer. He had been Otto and Langen's Parisian representative in the 1870s and became a personal friend of Daimler. He was subsequently the agent for Daimler's engine sales in France, assisted him with patenting and had coincidentally obtained the French patents for Daimler's engines. Sarazin also linked Daimler, Levassor and Armand Peugeot who soon obtained the licence to manufacture Otto and Langen's *Deutz* gas-fuelled engines which they held until the advent of Otto and Langen's *Silent* engine in 1878 (Chapter 14b).

When Périn died in 1887, Panhard and Levassor began making IC cars under the Panhard-Levassor name. Sarazin had also died in 1887, with Levassor at his bedside. The dying man is said to have urged his widow "*Keep working with Daimler – nobody today has any conception of the enormous potential of his patents*." They saw Daimler's *Wire-Wheel* car (Fig. 97) at the Paris Exhibition in 1889, signed a licence agreement with Daimler and kept the car in Paris. Panhard-Levassor quickly adopted Daimler's new Phoenix engine and their first car – powered by a Phoenix engine – appeared in 1890. The process was aided by the fact that in the same year Levassor had married Sarazin's widow, who had inherited Daimler's French patents and the role of acting as Daimler's French agent.

The Peugeot family firm operated three large tool-making factories in Valentigney in the Montebéliard region of France south-east of Paris. Under Armand Peugeot's[716] leadership the firm was also producing cycles and steam-powered vehicles. Montebéliard was a few hundred kilometres south of Mannheim and in 1888 Daimler sent one of his engines by railway to Valentigny. Although committed to steam engines, Armand

Peugeot quickly and perceptively recognised the advantages of the lighter DMG IC engine and became its first purchaser.

In 1891 the Panhard, Levassor, Peugeot trio were able to drive a DMG-powered Panhard-Levassor over the 400 km between Paris and the Peugeot factory in Valentigny. It was the first designed-for-the-purpose car (Fig. 98), with the engine placed over the steered wheels at the front of the car but with the power transmitted to the rear axle in the manner described below and shown in Fig. 96. They sold six of the vehicles in this first year. The vehicles were initially powered by the Daimler two-cylinder engines, however one advantage of the Panhard-Levassor arrangement was there was no practical limit on engine size.

Fig. 98a An 1896 Panhard-Levassor (see also Fig. 98b). via Wikipedia Commons.

Sprocket and chain drives had been widely used in association with the stationary gas-fuelled engines and the process of gearing by providing a means of selecting the appropriate pair of sprocket wheels was commonplace. Levassor made a major and lasting contribution to motor vehicle operation in switching to using a shaft to transmit power from the engine to a gearbox which used bevelled gears within an enclosed casing to allow a range of speed/power combinations to be selected whilst power was interrupted using a friction-operated clutch (Fig. 98b). A pair of chains then took the power from the gearbox to the driven axle. In 1891-8 Louis Renault[717] replaced these chains with a single shaft to the rear axle, and produced a gearbox with three forward speeds and one reverse speed. The arrangement was to persist for another century. Renault had worked for the Serpollet steam vehicle company (Chapter 13b) before beginning to produce his own cars in 1897 when only 20 years of age.

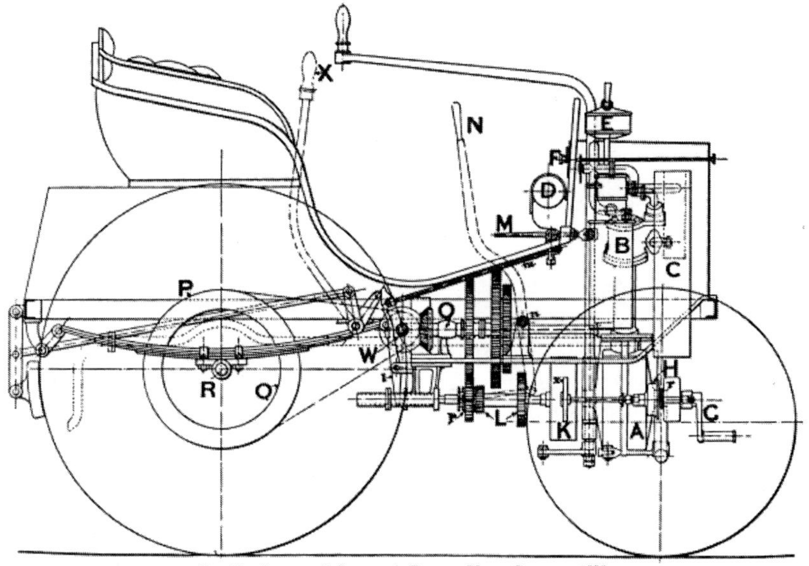

FIG. 56.—PANHARD & LEVASSOR'S DAIMLER MOTOR CARRIAGE, 1894.

Fig. 98b Panhard-Levassor's new power transmission system. The vehicle itself is shown in Fig. 98a. From [718].

By the end of 1891, the world had 13 IC cars – seven powered by Benz engines but built in France and six produced by Panhard-Levassor, also in France. There was also at least one Peugeot prototype from 1890 and in 1892 Peugeot produced 29 IC cars which were very similar to the Daimler *Wire-wheel* car (Fig. 94). Thus, despite their German engines and chassis design, the new IC cars were predominantly built, promoted, operated and enjoyed in France.

In France, car production continued to increase and cars were a common sight on Parisian streets in 1895. At the same time the first advertisements for cars appeared and the Automobile Club de France was formed and quickly attracted over 400 wealthy members. The word "*automobile*" had been coined by the French Academy in 1875 to describe steam-powered buses. It was adopted in the USA as a replacement for "*horseless-carriage*". Taking "*car*" from "*carriage*", the English initially used "autocar", but this soon became "*car*". This book does not use "*motor car*" as "*motor*" is redundant due to the word *car* now implying a self-powered vehicle.

All these French developments positively affected DMG IC engine sales and so in 1895 Daimler and Maybach were induced to return to DMG, with considerable encouragement from Frederick Simms,[719] an English engineer whose family had a business in Hamburg. He had become friendly with Daimler and was a great admirer of his engines. In 1893 he was made a director of DMG and formed the Daimler Motor Syndicate in England. This firm became the Daimler Motor Company in 1896 and initially made Daimler engines for boats. By 1897 they were producing cars in England powered by Daimler-patent engines.

The return of Maybach to DMG saw a new four-cylinder (third) version of DMG's Phoenix IC engine. It was one of Maybach's many exceptional and unparalleled contributions to the development of IC engines. The French car manufacturers were quick to adopt the new engine. In 1895 DMG also began producing a car of its own carriage-like design known as the *Riemenwagen* (Fig. 99). In 1898 DMG introduced the 4 kW *Victoria Wagonette*[720] with an improved gearbox and in 1899 it produced a racing car with an 18 kW, 4 cylinder Phoenix engine.

Fig. 99 DMG's horse-carriage style *Riemenwagen*, 1897 model. Note the steering and pivoted front axle. Photo with kind permission of the Mercedes-Benz Classic Archive (U75473A).

The fact that the first designed-for-the-purpose car – the Panhard-Levassor (Fig. 98) – came out of France rather than Germany did not surprise one major commentator who observed that the two major German engine producers had vacillated in opposite directions over the most appropriate vehicle design.[721] Benz had begun with bicycle-based design – the *Patent Motorwagen* (Fig. 93) – and then moved to horse-carriage-based *Viktoria* (Fig. 95b) and in 1894 returned towards the bicycle with the *Velo* (Fig. 96); whereas Daimler had moved from the carriage-based *Motorkutsche* (Fig. 94) to the bicycle-based *Wire-wheeled* car (Fig. 97) and six years later in 1895 back to the carriage-style *Riemenwagen* (Fig. 99). The situation was heightened by Benz' lack of interest in big, fast cars and Daimler's view that road vehicles would not be his major market.

However, in France there was strong interest in smart, fast cars and so, during this period of German vacillation, the new Panhard-Levassor layout described above was developed and set a new market standard. It came to be known as the *système Panhard* and has remained as the common car arrangement, although the initial versions (e.g. Fig. 95) still used tiller steering.

In the 1890s another German engineer, Rudolf Diesel[722] working in Berlin, was concerned with the low efficiency of existing IC engines. He was greatly influenced by Carnot's theoretical work on the efficiency of heat engines (Chapter 7b) and used its principles to design an engine in which high compression in the cylinder raised the air pressure sufficiently to heat the air to a temperature that would ignite fuel which was then injected into the cylinder. In the previous engines the fuel was mixed with the air being sucked into the cylinder.

Diesel's first engine was produced in 1893. His engines were relatively large but, by the end of the century, were in wide use for applications – such as ships –where high power outputs were needed. Today, diesel engines dominate the market for high-powered IC engines and have about half the car market.

CHAPTER 16C

INTERNAL-COMBUSTION HITS THE ROAD

In the last half-decade of the 19th century, the IC-powered vehicle was beginning to make its presence felt, although there are no indications that anyone foretold the impact it would have on all aspects of the following century.

The genie probably emerged from its bottle in 1894 during a competition for self-powered vehicles ("horseless carriages") over a 127 km route between Paris and Rouen (*Concours du 'Petit Journal': Les Voitures sans Chevaux*). Although a steam-powered De Dion-Bouton vehicle (Fig. 80) recorded the fastest time, it required both a driver and a stoker and so the judges announced that the two winning entries on the basis of ease of use were a Panhard-Levassor and a Peugeot with their two-cylinder DMG IC engines.

A year after the 1894 competition, an actual road race was run over a 1200 km route from Paris to Bordeaux and back. The 21 vehicles in the race comprised fifteen IC, six steam and one electric vehicle. Only nine completed the race. All of these were IC-powered – six used DMG engines, two used Benz engines, and the other was Léon Bollée's[723] *Voiturette*. The *Voiturette* was a three-wheeler with a DMG-type engine although its single cylinder 2 kW engine was mounted horizontally beside the single rear wheel.[724]

Émile Levassor famously dominated the race by driving one of his Phoenix-powered Panhard-Levassors for 49 h at an average speed of 25 km/h. He was so far ahead of his competitors that he stopped for sandwiches and champagne before crossing the finishing line. The event gained great publicity for IC cars and sales catalogues were eagerly sought. Levassor died in the following year in a car-race crash caused by swerving to miss a dog.

Levassor was passionate in promoting the new cars and his endeavours, combined with German indifference, and the fact that France had better roads and a more effective petroleum distribution network all meant that car production and marketing at the end of the century had moved predominantly to France. It has also been suggested that the reason for the better availability of petroleum products in France was that France had a relatively small coal industry, whereas many of the finance houses that controlled the production and distribution of petroleum were under French control and would gain if energy production moved from coal to petroleum. Indeed, Beasley suggests that, just as the British coal industry had actively supported the widespread introduction of coal-burning steam railways in Britain (Chapter 11e), so the French oil interests half a century later had actively encouraged the public availability of internal-combustion road vehicles.[725] McShane treats this view sceptically, noting that although American oil barons such as the Rockefellers had personally enjoyed using the new IC cars, they had made no early effort promote or invest in the embryonic industry.[726] It certainly assumes that the French investors had exceptional and unique foresight in envisaging a wide public market for IC road vehicles.

By 1898 DMG was selling four-cylinder versions of the Phoenix which had a power output of about 6 kW (or twice that of the previous 2 cylinder Phoenix) although it still had a number of relatively primitive features such as hot-tube ignition. DMG also used it in a vehicle of its own design and manufacture – the *Cannstatt-Daimler* (Fig. 100). In style, it was a significant advance on the *Riemenwagen* (Fig. 99). Following the Panhard lead (Fig. 95), the engine had been moved to the front of the vehicle. Mechanically, the vehicle had screw-down brakes which were common in horse-drawn carriages and in steam-powered locomotives prior to the 1870s (Chapter 11f). They were more for holding a stationary vehicle in position rather than for stopping it and are indicative of the general low-speed travel environment.

Fig. 100 *Cannstatt-Daimler* car, with its Phoenix engine providing about 6 kW of power, c1898. Photo with kind permission of the Mercedes-Benz Classic Archive (U746).

The 1899 Cannstatt-Daimler had further advances, including a 3 L four-cylinder engine producing 8 kW. The associated Phoenix racing car had a 5 L four-cylinder engine that operated at 160 Hz and produced 20 kW of power. Maybach's design input was considerable, particularly his pioneering invention of the spray-nozzle carburettor which would become the model for all carburettors for decades to come. Maybach was now running the company, as Daimler was succumbing to the illness which was to lead to his death in 1900. One marketing problem with the Daimler engines was that they still used a heated tube – rather than an electric spark – to ignite the fuel mix and this could make them difficult and slow to start.

Emil Jellinek[727] was a German-born banker who had developed a passion for the new IC-powered cars whilst living in Nice. He was a competitive driver and won the first Tour de Nice, driving as *Mr Mercédès*, which was his daughter's name and meant "*grace*" or "*god send*" in Spain, where it was common given name. In 1893 Jellinek had visited Daimler in Cannstatt seeking and demanding faster cars. He subsequently bought and raced a number of DMG vehicles. Jellinek then moved to Paris where he became one of the first car-dealers.

After a fatality on a test track in early 1900, and soon after Gottlieb Daimler's death, Maybach and Jellinek decided that the DMG car needed some fundamental changes. The new car had an improved Phoenix engine producing 26 kW, a body with a longer wheelbase to better distribute the engine mass between axles, and a lower centre of mass (Figs. 99 & 100). Within the year Jellinek had ordered 36 of the redesigned cars from DMG. However the name Daimler was no longer owned by DMG (Chapter 16a) and so, when the first car of Jellinek's order was produced in 1901, Jellinek requested that it be named *Mercedes*. DMG delivered the cars, gave Jellinek exclusive sales rights in France, Austria and the USA, and made him a director of DMG.[728] The 1901 *Mercedes* was the first modern car in all essentials, as can be seen by comparing Figs 101 with Figs 97 - 100. It was about 5 times more expensive but much more user-oriented than the *Velo* (Fig. 96), which was still being marketed by the Benz company.

In 1902 DMG adopted the name *Mercedes* for most of its cars. When DMG merged with the Benz company in 1926, their products were henceforth called *Mercedes-Benz*.

Fig. 101a The first *Mercedes*, built in 1901. It was the first modern car in all essentials. See the racing version in Fig. 101b. Photo with kind permission of the Mercedes-Benz Classic Archive (22381).

Fig. 101b The racing car version of the first *Mercedes* Photo with kind permission of the Mercedes-Benz Classic Archive (25074).

In 1893 de Dion (Chapter 13b) had pressed one of his two partners in steam engine production, Bouton, to agree to de Dion-Bouton producing an IC engine which would compete with the Daimler engines being used by their competitors. He succeeded in his argument and his single cylinder engine was probably the best on the market. The initial version only weighed 20 kg and was much lighter than other engines. It ran at a much higher engine speed of 30 Hz, producing about 300 W. Later versions could reach 60 Hz, with a power output of about 3 kW. Meanwhile, the third partner, Trépardoux, resigned from the firm in protest over the move from steam to IC. In about 1895 de Dion-Bouton switched all production to IC engines. De Dion's father tried unsuccessfully to prevent this move, telling a court that his son's predilection for mechanical vehicles was "*childish and scarcely admissible for a normal man*".[729]

Another epoch French car race was conducted in 1896 over some 1800 km from Paris to Marseilles and back. The first three places were won by Panhard-Levassor vehicles, but one of the relatively small De Dion three-wheelers was the first vehicle to reach Marseilles and three De Dions completed the race. This gained the car great publicity and a corresponding large increase in sales as the low-powered (c1.5 kW) vehicles were relatively cheap and easy to drive and maintain. In this respect, there were strong similarities with the market impact of the small Benz *Velo* vehicles.

In 1895 there were 350 self-powered vehicles in use in France and 75 in Germany. Most had been made in France using the German engines. The largest supplier was Benz & Cie with 135. Panhard-Levassor and Peugeot had each produced 75, mainly with the Daimler/Maybach Phoenix engines. Panhard-Levassor soon became market leaders in terms of price and quality. By 1898, Benz had sold 1130 cars including 510 in France and 330 in Germany. In that same single year France produced 1500 cars. By 1899 there were 75 different car manufacturers in France. The ascendancy of French car manufacture is further indicated by the fact that, as the century ended, Benz had produced about 2000 cars whereas de Dion-Bouton had sold about 10 000 and had become the world's first high-volume car producer.

Daimler and Benz had both been lured to France by a number of very tempting factors, particularly the ready availability of finance and of better manufacturing capabilities linked to a hungry market, the absence of government regulations inhibiting the use of powered vehicles and better roads due in part to a French law of 1836 which had made it compulsory for local government to maintain roads and provided a taxation stream to fund the work.[730] Perhaps the citizens of south-western Germany were also more attune to Parisian attitudes than they were to those of Prussian Berlin? They must have been considered decisions as Berlin had long been a centre of coach-building (see Chapters 1d&e) and as many German cities were active in building steam-powered locomotives (see Chapters 14b&c).

Before the end of the century, IC-powered cars were also being produced in Italy (by Enrico Bernadi[731] from 1890), USA (by the Duryeas[732] from 1893) and in England (by Knight[733] from 1895 and Lanchester[734] from 1897).[735] America was then producing about 3000 vehicles per year. Back in Germany, Daimler and DMG had produced less than a dozen cars by the end of the century. As mentioned in Chapters 14b and 16b, Daimler never saw the car as an interesting or important market. Others did see the market potential and by 1904 America was producing more cars than Europe.[736]

The absence of any British players in the car industry is sometimes thought to be entirely due to the restraining effects of the Red Flag Act (Chapter 13a). However, a past concentration on steam locomotives at one extreme and bicycles at the other, had impeded the development of a manufacturing industry able to match the quality and quantity of cars being produced in France. At the turn of the century many in Britain still considered the new cars to be toys, a view confirmed when only 10 of the 33 entrants in 1896 London-to-Brighton *Emancipation Run* actually finished the course – "emancipation" referred to the removal of the Red Flag Act in 1896. Compare this poor result with the performances of vehicles in the much longer French "runs" held over the previous decade (Chapters 13b and 16c). The situation was worsened by Henry Lawson (see Chapter 2 for his previous role in the bicycle industry) who used his ownership of Daimler patents and French vehicle manufacturing rights to inhibit the development of a British car industry.[737]

It is perhaps fitting to end this part of the story of 19th century transport to note that, in the last decade of the century, major European financiers led by the French arm of the Rothschild family, moved into the car industry, initially buying the de Dion-Bouton firm and becoming strong promoters of all aspects of "*automobilism*" in France. For the wealthy, automobilism provided an amusing and wonderful new toy and was the epitome of conspicuous consumption. These early drivers were certainly widely disliked by the general public and the drivers themselves often treated the public with disdain and arrogance as epitomised by the de Dion advertisement from 1899 where the vehicle occupants are paying no attention to their effect on other road users (Fig. 102).

Drivers sometimes declared that any damage caused to life or property was not their fault but a consequence solely of the inferior roads that they were forced to use. Another attitude was displayed by Rudolf Diesel (Chapter 16b) who was an avid long-distance motorist. In his 1905 diary he describes how he was travelling down an Italian valley and looked back to see a scene of astonishing beauty – his car had filled the whole valley with a wondrous white cloud of swirling dust:[738]

Fig. 102 Wilhio's de Dion-Bouton advertisement from 1899 with car occupants oblivious to other road users.

> *What a dust storm we stirred up leaving Italy! Powdery limestone dust lay 50 mm thick on the street. Georg raced along, demanding from the car all it had to give, through the Piave Valley, and behind us there swelled a colossal cone. This white cone rose in the air and expanded to infinite proportions. The entire Piave Valley was thick with the fog, a white cloud lying over the valley all the way to the mountain ridge. We outraged the pedestrians with gas attack – their faces pulled into a single grimace – and we left them behind in a world without definition, in which the fields and the trees in the distance had all lost their colour to a dry layer of powder.*

Late in the 19th century the oil industry also had a strong reason to take an interest in the new vehicles powered by IC using petrol as a fuel. Extensive oil fields were still being discovered – for example in Romania and Russia – and larger refineries were being built closer to markets. However, rapid developments in the production and distribution of electricity were taking away markets that had been oil's initial customers, particularly the change from lighting by kerosene-burning lamps to incandescent electric lamps, which became practical following the work of Thomas Edison at Menlo Park in New Jersey, USA, in 1879. There were signs of a glut occurring in the oil market and major new consumers were needed. Perhaps the IC engine would be the answer to the oil maiden's prayer?

There was as yet no hint of the mass market that Henry Ford was to tap with his *Model T* in the first decade of the new century. However the signs were there, even for the wealthy French owners of the new cars. In 1898, whilst a car cost 20% more than the equivalent horse and carriage, its running costs were only a third of the horse and carriage running costs. The new cars were providing far better service at half the total operating cost of the horse and carriage.[739] The annual running cost of a horse and carriage exceeded its purchase price; the opposite was true for a car. Furthermore, as discussed in Chapter 1e, each horse required about 2 hectare of land to supply its food needs. Apart from these mechanical advantages, the IC cars also liberated travellers from the numerous corporal demands of the horse – such as concern for its physical well-being, its variable emotional state, its needs for rest and attention and for fuelling even when not in use, its frequent need to defecate and urinate, its potential to cause harm to bystanders, and the reality that – if damaged – it was almost impossible to repair.[740]

The fact that the car had yet to appeal to the mass market meant that its dramatic effect on the way the world works had yet to be felt when the 19th century ended. The car was still predominantly a plaything being used for social rather than for business and journey-to-work applications. However a leading New York commentator editorialised[741] in 1899 that "*the experimental plaything has become a practical necessity*." The IC truck at the same time was strongly catching the eye of the military officers but had yet to appeal to commercial logistics managers.

It is nevertheless appropriate to remind readers of the unique characteristics of the IC-powered car which led to its 20th century dominance. Colin Buchanan was one of the 20th century's great commentators on the influence of the car and it is appropriate therefore to use his mid-20th century words to convey the essence of the car's uniqueness:[742]

> "*In particular it may be questioned whether the fundamental nature of the transport revolution has been grasped, whether it is appreciated that the*

advent of a means of personal mechanical locomotion adaptable to a host of everyday journeys has, by turning the streets into rivers of jostling, lethal vehicles, rendered out of date at a stroke the conventional arrangement of streets and buildings that has served us for so long. If the motor habit has its heyday yet before it....."

A few decades later another respected commentator, Clay McShane[743] made a slightly different observation, noting that in the 1890s and preceding the arrival of the car, "*American urban culture changed.* City *residents perceived their streets in a different way, thinking of them more as trafficways than open public spaces*." Of course, it has already been shown in Chapter 12b that many urban streets in both the Old and New Worlds were congested "*trafficways*" for most of the 19th century.

The first road built specifically to match the capabilities of the motor vehicle was the Long Island Motor Parkway built in New York State in 1906 by William Vanderbilt, to provide a road on which he and his wealthy friends could drive at their highest desired speeds.[744] Ironically, Vanderbilt had made his fortune from the American railroads. The new cars, although seen as toy for the wealthy, would soon offer no respite from this urban contagion. McShane notes that there was no grand design by the technologists that led into the vast markets of the 20th century. Instead there was a coalescence of invention and of many social factors. Was this coalescence accidental or the inevitable outcome of the preceding century? It was certainly not a perfect outcome but, it is inconceivable that mere-mortal decision-makers could have had the vision to see where the changes were taking them. The lesson to be learnt is not that we need to seek decision-makers with super-human powers of foreseeing the future, but that we always need to have a strong system of checks and balances that will preserve and maintain those things that we think are important, come what may.

The story of the car from 1900 on is well known[745] and is no longer of direct relevance to the theme of this review. However, it should be said that today's vehicles represent a long series of incremental improvements over the first vehicles produced by Benz and by Daimler and his associates. This period of over 120 years flies in the face of the well-known view advocated by Nickolai Krondatieff in 1925 who suggested that most major capital goods expansions occur over a period of about 60 years and which was still being propounded in 1990 by Rostow,[746] despite the long track records of Mercedes-Benz, Peugeot, Ford and General Motors.

PART 17

PAVING THE WAY

CHAPTER 17A

THE BRITISH ROAD SCENE AT THE END OF THE 19TH CENTURY

The influence that the better French road system had on luring, not only Benz and Daimler, but also the whole car-manufacturing sector to France towards the end of the 19th century was discussed in Chapter 16b. How had this happened, given the economic dominance of Britain for most of the century and its excellent coaching network in the first quarter of the century?

Clearly, the commercial success of the British railways and all the associated industries, as described in Chapter 11, had focussed British commercial and social attention on the railways. Add that success to the "failure" of Telford's Great Irish Road in the 1840s (Chapter 4c), and it is not surprising that the British Government saw no need for road-based initiatives in the second half of the century.

At the height of the coaching era in the 1830s Britain had some 15,000 separate authorities administering its road system. By the end of the century, despite dramatic reductions, there were still 1,800 responsible authorities.[747] There had been some continuing mild pressures for a central road administration, but it did not fit the very local pressures dominating the British Parliament's occasional consideration of road matters. This was despite the obvious success of the centralised system used in France and described in Chapter 3b. Even when the British Parliament did finally direct its attention to road administration and maintenance in 1878, it created regional highway districts governed by local boards, with no suggestion of central control. Indeed, from 1872 to 1919, there was no British Parliamentary Minister answerable to Parliament for anything related to roads, although as much money was being devoted to roads as was being spent on public education.[748]

The prime local reason for this situation was that it was generally accepted in most communities that steam railways were the obvious transport system for other than local trips and that the roads and the associated steam- and horse-powered vehicles would best serve local needs or provide feeder services to the railway stations. Plowden[749] explores this in some detail and I have recorded a similar example in my home Australian State of Victoria.[750] There was no need for a wider view of the road system. The century had passed before the incorrectness of this common view began to dawn on most people, irrespective of their political or social outlook. Guldi has provided an alternative explanation of these developments, based on social and political arguments and applying labels such as *radicals* and *libertarians*.[751] I have little knowledge or understanding of these factors, but I do feel that the events in Britain – and in many other places such as Victoria – can be fully explained by the daily workings of society without recourse to Guldi's arguments.

New pressures to improve British roads in the 1880s came from three main sources:

1. The realisation by the railways that they needed roads as feeders to their system (Chapter 12),
2. The steady improvement in steam traction engines (Chapter 13b), and
3. The increasing public popularity of bicycles (Chapter 2).

The arrival at the end of the century of vehicles powered by internal-combustion engines was to provide an overwhelming fourth pressure, but being of the 20th century it is beyond the scope of this book.

As a result of these pressures, common sense began to prevail and a British Act in 1888 vested control of all main roads in the county councils and in 1894 control of the remaining roads was vested in rural district councils.[752] There was no perceived need for a national overview as the railways adequately filled that need. Thus the final developments in Britain in the last decade of the century at last had a semblance of logic and reasonableness – although the advent of the motor vehicle quickly destroyed that logic and reason. As in most other countries, decades of the 20th century were to pass before road planning and management began to accept the realities of the motor vehicle.[753]

CHAPTER 17B

EMERGING TECHNICAL INADEQUACIES OF 19TH CENTURY ROADS

The better understandings of pavement technology introduced by McAdam and Telford and the development of asphalt and concrete (Chapter 17c) meant that there were some, albeit poorly funded, improvements in road condition during the 19th century. There had also been major work aimed at reconfiguring the street networks of Paris and London (Chapter 12b). However, despite the early work of Telford and Macneill on optimum road geometry (Chapter 4c) and some French roads widened to suit Napoléon's military needs, during the 19th century there were only minor geometric alterations to most roads in terms of gradients, curve radii and pavement widths.

The importance of this last observation is that at the end of the 19th century legal speed limits had risen to 20 km/h in France, the home of most of the new cars (Chapter 16c). The change reflected reality for in 1895 in the first conventional car race which ran from Paris to Bordeaux and back (Chapter 16), Émil Levassor's average speed in his Panhard-Levassor was 25 km/h. This was soon prove to be a minor step for just 5 years later a steam-powered car had travelled at over 100 km/h.[754]

Chapter 1e described how, when the road system was at its peak in the 1830s, Royal Mail coaches in England were operating at speeds of 25 km/h. Many speed-related effects increase with speed squared, so a four-fold speed increase suggests a 16-fold increase in consequences. However the roads themselves were unaltered. Observing the situation in the USA "*just before 1900*" a leading motorway advocate observed that "*what had been good enough for the automobile's grandfather was considered good enough for it*".[755] Thus it should be enlightening to explore these effects of this short-sightedness in more detail.

Chapter 10b discussed how human response times were inadequate for people confronted with the new trains in the 1840s. The cars on public roads were travelling at twice the speed of the first trains on their restricted right of way. It is useful to recall that 100 km/h is about 30 m/s, so – if a total human response time is about 2 seconds – 60 m of travel will have occurred between observation and immediate consequence.[756] If a car driver observes an obstacle ahead that could lead to a collision, a common driver reaction would involve braking. Even with today's brakes, the effective stopping distance for a vehicle initially traveling at 100 km/h, is about 200 m.[757] Thus, a driver travelling at 100 km/h needs a clear view ahead of about 200 m. The technical term for this "*clear view*" is *sight distance*. Sight distance is reduced by curves, crests and low light levels. Right angle corners provide hardly any sight distance. Obviously, many 19th century roads did not provide the sight distance needed for the safe operation of the new vehicles being produced at the end of the century. As a consequence, the roads the new century inherited from the 19th century were grossly inadequate for the new vehicles that came as part of that same inheritance.

An early advantage of the railways was that they were built on new alignments chosen to permit high-speed travel whereas the road system operated on alignments created by a series of historical and geographic accidents and irrelevancies. Furthermore, whereas the railways operated on smooth tracks provided by the iron rails, road surfaces were variable, unpredictable and easily influenced by the weather. There were relatively few new road alignments created during the 19th century, but road-building technology did improve and paving technology improved dramatically.

Thus, when turn-of-the-century regulators in most developed countries allowed vehicles to travel at speeds of 100 km/h or more on the existing road systems with their poor alignments and surfaces, it must be charitably assumed that they considered that vehicles travelling at such speeds in such conditions would remain relatively rare. The unfortunate legacy of such high permissible operating speeds had many undesirable effects in 20th century.[758]

There was also a curious trichotomy amongst the car-makers who made vehicles capable of travelling at well over 100 km/h, yet had their maximum mechanical efficiency at speeds closer to 80 km/h, and would spend much of their working lives at speeds of 60 km/h or less. Thus the 19th century ended with no signs of any optimisation of the traveller-vehicle-road system and the new century accepted this situation as established normality.

Chapter 4a listed road width as one of the four factors determining the effectiveness of a road. Initially, paths and roads would have been wide enough to accommodate a single user, perhaps with occasional provision for stopping and passing. Human widths have altered little over the millennia, so path widths have been mostly in the 500 mm to 800 mm range. Vehicle widths of about 2 m have also been relatively constant over time, due to the need to allow for the width of the hauling animals, harnesses and shafts. This determined a wheel spacing of about 1.4 m (Chapter 9c). Consequently, a common historical roadway width has been about 2.5 m.[759] Although modern vehicles are much faster, they are also easier to control and so lane widths today for very light traffic are as narrow as 3 m.[760]

Over time therefore, road widths have been determined by how many lanes of traffic are to be accommodated, the need for shoulders for emergency stopping, the width of roadside drains, and any side clearance required for additional safety or operational reasons. The 1835 British Highway Act required roadways to be 6 m wide, which would have provided for one lane traffic in each direction. There were exceptions; early in the century Napoléon had required French main roads to have a paved width of 7 m in order to accommodate the marching movements of the French Army. On the road to Holyhead (Chapter 11a) Telford added a 2 m shoulder on either side, resulting in a roadway 8 m wide. By the end of the 19th century, a leading road engineer was advocating a width of no more than 12 m for major roads.[761] Such roads could carry two lanes of traffic in each direction, with the occasional prospect of overtaking or passing. Once a road right-of-way reservation is established, property development often occurs along the edge of the reservation, making future expansion of the road a difficult task. Thus narrow roads resulting from an under-estimation of future road traffic were an unfortunate legacy that the 20th century received from its predecessor.

In cities a different situation prevailed. In most cases the urban inheritance of previous centuries had been a city bounded by protective walls and was commonly less than 1.4 kHa in area. The streets were usually narrow to avoid wasting valuable land and as there was no need to provide for through traffic or for many wheeled vehicles. Entrances to the city were restricted to bridges and to gates in the city walls. Catering for traffic was rarely an administrative imperative. Thus, even before the car, most urban streets were inherently congested.

Traffic congestion arises from two main causes: when demand for road space exceeds supply, or when chaotic conditions initiate local blockages, which then cause system-wide impacts.[762] Congestion on roads at the beginning of the 19th century was discussed in Chapter 12b. Demand exceeding supply congestion due to motor vehicles became a major 20th century issue and will not be discussed further in this book.

Road safety had been a problem prior to the 19th century. Animal-drawn vehicles were always a potential threat to life and limb if the vehicle rolled, or if pedestrians became entangled with the wheels, or if the horse bolted. The problems were exacerbated in the 19th century as lighter coaches and fitter horses permitted faster travel speeds (Chapter 1e). Indeed, some data suggests that urban traffic fatalities per distance travelled were greater in the 19th century than in the early 20th century.[763]

The first car-related fatality occurred in 1895 when Émile Levassor died in a crash in a car race (Chapter 12c). When Benz & Cie issued the world's first car advertisement in 1888 (Fig. 103), it proclaimed that its patented car was "*comfortable and totally safe*" and, indeed, 12 years were to pass before the first road death involving a car on a public street. This tragic event occurred in South London in August 1896 when Mrs Bridget Driscoll, who was crossing the road at the time, was struck by a passing car. The driver had seen Mrs Driscoll and had shouted, "*Stand back*" and rung his bell quite loudly. What more could he have done? The coroner expressed the clear wish that such an event would never be repeated.[764] The car, incidentally, was a Roger-Benz (or Model 3, Chapter 16a), which was about three times as powerful as the model in the advertisement.

Rules such as travelling on one particular side of the road did apply on most 19th century roads[765] but quickly broke down in congested town and city traffic. Such 19th century traffic congestion was often caused by chaotic situations (Figs 104 & 105). What rules governed travellers fighting for road space? How was a herd of animals bound for a market to be managed? If a vehicle was very large, or its haulage animals frightening, or its occupants very important and/or well-armed the might-is-right principle would apply, although the awkwardness of large vehicles could cause another set of problems. Otherwise, the underlying problem was often the lack of any definition of priority. Subsequently, chaotic congestion has been well managed by introducing traffic regulations which basically define who gives way to whom. They work surprisingly well but were not widely adopted until after the First World War,[766] although New York had introduced a comprehensive traffic code directed at horse-drawn vehicles in 1897.[767]

The general issue of driving on the left or right raised in Fig. 104 has no relevance to this book and is well covered elsewhere.[768]

Two other factors that made 19th century congestion worse than our current perceptions are that vehicles of that time occupied far more road space per traveller or per ton of freight than current vehicles and they travelled at less than 20 km/h and so traffic throughput was limited. If the traffic speed is halved the traffic throughput is halved. So to accomplish the same transport task, the 19th century required a larger transport fleet than is needed today. The situation is a little more complicated than this in today's traffic, as slower vehicles require smaller inter-vehicle headways.[769]

Fig. 103 The first car advertisement, issued by Benz & Cie. in 1888 for Model III of their Patent-Motorwagen. See also Fig. 93. The words at the top read:

Patented in all industrial countries!

New! Practical!

Gas-operated by petrol, benzene, naphtha, etc.

Always starts immediately! - Comfortable and totally safe!

Part of photo with kind permission of the Mercedes-Benz Classic Archive (22588).

Fig. 104 Chaotic London traffic congestion in the Temple-Tower area late in the 19th century, as seen by Gustav Doré in 1872.[770] Note that there is no clear lateral distinction between travellers moving in opposite directions – are the vehicles keeping to the left or to the right?

Fig. 105 Chaotic traffic congestion on Broadway, NY, in 1883. From McShane[771].

CHAPTER 17C

ROAD PAVEMENTS IN THE LATTER STAGES OF THE 19TH CENTURY

The demand for better paving, particularly in the cities, had arisen well in advance of the IC-powered car, although the new cars and trucks quickly took advantage of good pavement surfaces.[772] The demand had arisen as the benefits of the Industrial Revolution became more widespread and as cities became larger and more densely populated (Chapter 18a). In this context, well-paved streets were seen to be a key part of a liveable, vibrant city with major pedestrian flows (Chapter 12b). Of course, the need for good pavements with cleanable surfaces was exacerbated by the widespread use of defecating and urinating horses as a source of motive power (Chapter 13a) and enhanced by the increasing acceptance of underground pipe systems as a way to keep sewage and other wastes off street surfaces and well out of sight and smell.

In the face of these demands, a fortunate coincidence occurred with the rapid move described in Chapter 15a towards the use the "lighter" fractions of petroleum distillation – petrol and kerosene. The coincidence was that a very practical associated use was found for the otherwise unwanted "heavy" fraction of the distillation. The context of this coincidence was as follows.

Chapter 4a described how in the first half of the 19th century pavement construction was increasingly based on the revolutionary new method introduced and promulgated by McAdam. Subsequently, paving technology had improved more gradually.[773] The major initial improvement was the use of materials to fill the voids in the macadam stone layer and cement the stones together, particularly at the running surface. This process both water-proofed the pavement layer to prevent softening of supporting foundation and gave it tensile strength.[774] Three suitable cementing (or binding) materials arose during the Industrial Revolution – tar, bitumen and cement – although each had had limited use in earlier times.

Tar from burning coal and wood was widely available as a result of the many furnaces which had come into operation (Chapter 15a) and was a commonly used water-proofing material. From about the 1830s it was used in pavement construction, initially as a result of the serendipitous observation of the beneficial effects of a barrel of tar accidentally spilt on macadam pavements.[775] The resulting product in England was called tar macadam, which then was shortened to *tarmac*, a word which is still in common use.

Tar had some limitations in availability and quality (and much later was found to be carcinogenic) and so road-makers were always alert for alternatives. As noted in Chapter 15a, petroleum is a natural organic product and its properties vary dramatically from place to place. Sometimes it is comes to the surface as a flammable gas such as that which has fuelled sacred fires in Persian temples for over 5 millennia. At the other extreme petroleum is quarried with some difficulty as a rocky material created when petroleum softened by subsurface heat has infused into layers of broken surface rock. The mix then solidified as the oily petroleum remnant cooled. The resulting mix of rock and petroleum products was often called *asphalt*.

Between these two extremes is a range of petroleum components some of which are solid at ambient temperatures but become soft and mouldable at modest temperatures. Bitumen is one such product.

The gaseous and liquid components of petroleum are called "*light end*" products and thick oils and bitumen are "*heavy end*" products. Over the millennia, *heavy end* petroleum was obtained by heating asphaltic rocks. In service, the bitumen was heated to a malleable fluidity and then used for water-proofing, caulking cracks and joints, and covering major wounds – particularly amputations, battle injuries and castrations. Mixed as an emulsion with beer, it was taken to cure stomach ulcers.

Thus there was a market for bitumen and early in the 1800s there were a number of asphalt mines being operated to meet commercial needs. One of the large asphalt quarries was at Neuchâtel in Switzerland. There were other deposits being actively worked at Seyssel in France, Ragusa in Sicily, Limmer near Hanover in Germany and an entire lake in Trinidad.

In 1849 a Swiss engineer at Neuchâtel, André Mérian,[776] observed that the hard-rimmed wheels of carts carrying the rock to the heating vats were running over pieces of quarried asphaltic stone, grinding them into smaller particles and then forcing the particles into interstices in the existing surface, thus, by chance and with the possible aid of some hot weather producing a pavement better than any he had previously encountered. Recall that a similar explanation was offered for McAdam's invention of his broken-stone pavements (Chapter 4a). The potentially apocryphal nature of these to explanations is underlined by my finding of another 13 claims to have invented asphalt in similar circumstances.[777]

In asphalt, bitumen is the glue that holds the pieces of broken stone in place and helps provide a traffic surface that is smooth and water-tight. As in McAdam's pavements, the vertical loads from the wheels of the vehicles are carried by point-to-point contact between the pieces of stone.[778]

By the end of the century, asphalt (or similar cement) pavements were the norm and were providing high quality pavements of sufficient overall smoothness and local roughness to meet all the needs of the new motor vehicles. They were also much easier to clean and to drain. Petroleum and its offspring – petrol and bitumen – had provided the most convenient and effective of coincidences.

However, in contradistinction to the internal-combustion engine where development was pushed at a fast pace by brilliant inventors with little corporate baggage to restrain them, the development of asphalt pavements after the first flush of success was impeded by corporate restraints and the most modest of engineering inputs. The industry was driven by owners of the various asphalt deposits. Although the percentage and type of bitumen and properties of the stone that the bitumen had penetrated in the ancient past were acts of nature – indeed of a "nature" that had no particular interest in pavements – the corporate owners declared loudly that only their "accidental" stone and bitumen mix could make a good pavement. A bewildering range of product names was marketed and the technical development of asphalt was delayed for at least half a century. For most of the 20th century, asphalt pavement design still suffered from a heavy empirical bias and a lack of fundamental design parameters.[779]

Cement is the third of the major cementing materials developed during the Industrial Revolution. It was discovered accidentally by many early societies which had used pieces of limestone in the hearths of their fire places and had the observed the remarkable paste produced when they poured water on the burnt stone. This burnt limestone was mainly surface-reactive calcium oxide and silica di-oxide. Adding water produces various forms of calcium silicate which form a mouldable mortar that soon sets as a rock-like solid. The process was put to practical use by the Etruscans and the Romans.

Cement was reinvented in 1824 when Joseph Aspdin[780] was granted a patent for making "*a superior cement resembling Portland stone*." (Stone quarried at Portland in Dorset was then a popular British building material). At a site at Wakefield near Leeds, Aspdin burnt limestone in rotary kilns fired by coal or coke. His *portland cement* was a mixture of finely-ground burnt limestone and clay. When the fluid mortar was mixed with sand and stone it set to produce concrete, which became one of the most ubiquitous of the 19th century's construction materials. In particular, the innovative uses of reinforced concrete in structural engineering were discussed in Chapter 5d.

As mentioned in Chapter 4a, three other widely used paving materials in the second half of the 19th century were the traditional stone blocks (or setts), bricks made by pressing and heating clays and blocks sawn from wood.[781] They all suffered in comparison with the new asphalt and concrete pavements and were rarely being used by the century's end.

PART 18

PEOPLE

CHAPTER 18A

POPULATION EFFECTS

It might at first seem strange to devote a chapter to population issues in a book on transport. However people are the major consumers of transport, either directly or indirectly. In developed economies, as population in an area increases, so does the demand for transport. This is particularly reflected in the models used by transport planners to predict future transport needs.[782] Thus, as population is a significant cause of transport demand, population numbers and distribution are useful variables to describe and predict the extent of that demand.

Arguably formal transport planning began in 1837 when John Burgoyne,[783] then Chairman of the Irish Board of Works, was given the task of planning the Irish railway network. He did this very systematically by firstly measuring existing traffic flows on the road network, including vehicles used, passengers and cargo hauled, and their origins and destinations.[784] Based on the recent English experience with their new railways (Chapters 10 & 11), he could then estimate likely railway traffic.

In the absence of existing traffic, it was necessary to predict how many people at a particular origin would make a trip to a given destination. This task was first tackled by American economist Henry Carey[785] in 1857 when he recommended using an inverse square law – demand diminishes with the square of the distance between origin and destination – based by analogy on Newton's gravitation law. The associated planning process became known as the *gravity model* and, despite the unjustified bravado of its underlying assumption, was still in common use in the 21st century. A core first step in using these models is to predict the future population in an area. In practice, this has proved an almost impossible task and was rarely done with any semblance of accuracy, even in the 20th and 21st centuries.[786]

Population changes occur due to the effects of births and deaths and also due to permanent movements of people from place to place. The Industrial Revolution fell into the latter category for, as illustrated in Chapter 6, transport was just one of the many aspects of life that it affected in unexpected ways. For example, improvements in machine technology led to better farm machinery and in the 1840s this then led to major unemployment amongst the agricultural labourers who lived as tenants on the farms of the wealthy land-owners. Many had to move to the new industrial towns to find work. The insurgent working class and a slow decline in aristocratic power were to be major factors in British political life for the second half of the century.[787]

In England, some of those displaced and starving people then protested, following the precedent of the Luddite riots in the Midlands in 1811-2. Their main form of protest was to break the machinery that was taking their work. Consequently, they were known as "*machine breakers*". When caught they were harshly treated by the local magistrates and were often transported as convicts to Australia. Many others freely emigrated to the New World to seek a better life. Some, such as the author's paternal great-grandparents, were offered free passage from England in return for agreeing to work for a period on a designated New World farm owned by an Englishman.[788]

A specific example of the consequence of better machinery and more powerful engines was in the spinning machines which made cloth-making an industrial process, rather than a farming sideline. This created much rural unemployment and saw men in England moving to the industrial centres in the West in their search for gainful employment. Simultaneously, many women moved to London as domestic servants. Over the 19th century, the percentage of the British working population involved in agriculture dropped from about 40% to 10%.[789] The 1851 English census indicated that more people lived in towns than in the countryside. The imbalance continued for the remainder of the century and in 1901 only 22% of the English population lived in rural areas.

During the same period in which these changes were happening, Britain's total population was increasing rapidly – from about 9 million in the 1801 census to 16 million in the 1841 census, to 36 million in the 1901 census. The main initial contributor to the greater population growth was the increased life expectancy of children which had a major effect on population numbers in the last quarter of the century. It has been argued that previous population spurts had been due to factors that could not be sustained in the long term whereas the Industrial Revolution brought about changes which were sustained and whose effects were long-lasting.[790]

Thus, although some were gainfully employed in new industries created by the Revolution, others became unemployed, indeed unemployment was chronic. The latter factor allowed wages to be kept low and one major review[791] suggests that, at the half-century, wages for most workers remained at subsistence levels, although there was "*a slight improvement in average material standards...and most people were better off than their forerunners*". Clearly, the fruits of the Revolution were yet to be shared amongst all citizens.

Although living standards did rise in the second half of the century there were still pockets of appalling inhumanity, epitomised by the infamous Old Nicol slum on the edge of London's Bethnal Green parish (Fig. 106). In the 1880s the locality still had an annual mortality rate of 4% and a quarter of its live births died soon after birth. The locals were not recent immigrants, as the large majority of the inhabitants were at least second-generation Londoners. The seed of the problem had been planted at the beginning of the century when new manufacturing industry had moved into London and built some very inferior housing to accommodate its new workforce.[792]

Fig. 106 1863 Bethnal Green street scene. picture credit: Wellcome Collection. CC BY, via Wikipedia Commons.

The Old Nicol pattern is consistent with a pessimistic view of the Industrial Revolution. In 1901 an Oxford historian, Frederick Powell, said that "*the establishment of the factory system without the proper safeguards*" was the "*most deadly blow*" the English people had ever suffered. It was a commonly accepted view.[793]

At a broader level, prior to the 19th century cities had fluctuated in size as their power-base altered. London did not expand beyond its Roman walls until 1574. For most people, it is impractical to spend more than about an hour a day traveling to and from work and, in the era of pre-mechanised transport, this limited the size of cities to a walking radius of about 3 km. Population increases were managed by accommodating more people in the same area. By 1750 the population density of London had reached about 600 person/Ha, which was similar to the density in ancient Rome and Baghdad and appeared the maximum for cities that depended on walking for transport.

Rapid population growth occurred in London and Paris during the 19th century. For example, London's population increased from 1 million in 1801 (England's first official census), to 2.3 million in 1850 to 6.5 million in 1900. Population densities increased to about 700 person/ha.[794] The population growth was primarily accommodated by city expansion – thus in 1800 the clockwise ring of Knightsbridge, Buckingham Palace, Hyde Park, Paddington, Marylebone, Regents Park, Tottenham, Bethnal Green, Spitalfields, Tower Hamlets, Greenwich and Lambeth were all outside urban London. These areas – and many more such as Highgate and Hampstead – were embraced by the expanding city, usually with the aid of improved urban transport (see Figs 64 & 65 & Chapter 12a). The associated population growth of London during the 19th century has been well documented by Briggs.[795]

A common reaction to this situation in most of the world's cities was for the wealthy elite to build high quality housing in the urban core near their work-places. The poorer workers and various noxious industries were forced out to the new urban fringes.[796] The problem was exacerbated by the fact that the factories that arose during the Industrial Revolution were also commonly located on cheaper land on the urban fringes. Ironically, early in the 20th century the new cars reversed the process, allowing the wealthy to seek green open spaces on the urban fringes.

The flight to the cities as a consequence of the Industrial Revolution challenged the cities in many ways – transport was just one of those challenges. With no exception the cities of the world reacted to these changes very slowly and with limited foresight. Whilst this might seem a harsh criticism, the technological changes that did occur during the 19th century and the consequences of those changes –as this book has hopefully demonstrated – were far too great to have been foreseen and planned for by any mere mortal.

Many cities also had natural limits to geographic expansion and by 1900 New York had set a new population density record of 1900 person/ha, creating a whole new set of transport challenges for the next century.[797]

In summary, one of the main effects of the massive population increases that occurred during the 19th century was to place major public transport demands on the existing cities and on their urban fringes as the cities spread rapidly outwards. Understandably, this effect had not been foreseen and there were many undesirable consequences, some of which have lingered to the present day.

CHAPTER 18B

CHANGE-MAKERS

As the preceding Chapters have demonstrated, the 19th century was a time of many major advances in the way our world operates and most of these advances could readily be ascribed to particular individuals. This was a common 19th century view. Carlyle in his 1841 book[798] *On Heroes, Hero-Worship and the Heroic in History*, described how he saw history as having turned on the decisions of "*heroes*". He stated that "*The history of the world is but the biography of great men*", reflecting his belief that heroes shape history through both their personal attributes and divine inspiration. This heroic or Great Man approach to history was most fashionable with professional historians in the 19th century.

As in Chapter 19a, the Great Man theory is usually contrasted with one that talks about events occurring in the fullness of time, or when an overwhelming wave of smaller events causes certain developments to occur. Hegel in 1837 argued that a great man does not create historical reality himself but only uncovers the inevitable future.[799] One of the most forceful critics of Carlyle's formulation of the Great Man theory was Herbert Spencer, who believed that attributing historical events to the decisions of individuals was a hopelessly primitive, childish, and unscientific position.[800] He believed that the men Carlyle called "*great men*" were merely products of their social environment. This concept of the inevitability of a discovery occurring is often illustrated by the example of the simultaneous but independent discovery of calculus by Newton and Leibnitz in England and Germany in the 17th century.[801]

The story presented in this book largely aligns with Spencer's view. Men like Telford, the Stephensons, Faraday and Daimler had the exceptional skills, insights and persistence necessary to take advantage of the potentially fertile fields in which they found themselves. But were they unique in their attributes? My own interpretation is that they were not. They were at the right place at the right time to utilise their exceptional skills and they particularly stood out from their potential peers for their vision as to where their ideas would lead. They were not creating toys, they were creating new futures. But they were not God-like as none of the great contributors discussed in this book had any real perception of the future that would unfold as a consequence of their work. Their skills were in invention and the short-term market and certainly not in longer-term entrepreneurial foresight.

This puts my view of the 19th century on a very different plane to the widely discussed views of Joseph Schumpeter who argued in 1911 that foresight had a major role. In this much later review of 19th century transport, I have not found any evidence to support Schumpeter. However, the great contributors that I have discussed were all aware of the market place and of the then-current demands that might be profitably met by their inventions and break-throughs. There is also no 19th century evidence to support the more recent *"exogenous growth theory that sees invention occurring outside the economic system"*.[802]

I should state that I have not used the word "men" carelessly in this discussion. It would be nice to claim that my usage was gender neutral, however no women have played a role in the story I have told and 19th century societies would have made it very difficult for them to have done so. The possible exceptions, such as Marie Curie, worked in fields other than transport.

Pursuing the "*fertile fields*" theme I will now examine each of the contributors end-noted in the text with respect to their birth year and place. It could be assumed that the productive years of these contributors were between the ages of 20 and 60. Their birth places given in the end notes to this book may be good surrogates for education but should be viewed with care. Certainly, in writing the text it did seem to me that the type of education and initial work-experience a person received played a major role in their development. Even this conclusion must be drawn very carefully for, to take the two "heroes" mentioned above, Telford had a very basic education, self-educated himself to an astonishing degree and created markets for his exceptional skills. On the other hand, Daimler was carefully educated in an academic and a trade sense and found himself in the midst of technically active and creative industries. He was an exceptional and creative visionary, but before the century was out, his vision – although powerful and effective – was also found to be short-sighted. And Telford had also spent his last decades on an exceptional engineering scheme – the Great Irish Road – which was bypassed by new technology within a few short years. The lesson here is surely that, no matter how brilliant we might be at understanding, and inventing, and innovating – we are all dismal predictors of the future.

There are about 170 names in the end notes, but this reduced to about 150 when people such as Abraham Darby (Chapter 6) whose contributions were clearly before the 19th century are excluded. In considering these 150 individuals, the reader should apply four caveats. First, they are my subjective choices of the major contributors that arose as I – with English as my native language – was assembling the preceding Chapters. Second, some in the list had made many great contributions over decades whereas as others made just one spectacular contribution over a short period. Third, it is difficult to draw a line between transport and technology in general as the two go hand in hand. Fourth, and finally, the list is compiled from a very patchy data set. For example, Chapter 11f describes how locomotive technology in the second half of the century became increasing locked into the competitive activities of commercial entities and the associated and inevitable secrecy makes it difficult to detect specific advances and assign contributors to the advance. This applies not only to locomotives but to other areas, particularly in power generation and transmission.

Continuing this last theme, much of the relevant work involved the rising new profession of engineering. Whereas scientists were careful recorders of sources of ideas and of new understandings, making scientific advances relatively easy to trace and assess, no such discipline of a rigorous network of documentation developed in engineering. Organised documentation was via the very formal patent system which retrospectively staked precise claims to new intellectual land but did not describe the journey taken to reach that new territory. Nevertheless, the book draws heavily on patent data although, as mentioned in Chapter 7b, only Britain and the USA had strong patent systems throughout the 19th century. For much of Europe patenting only became effective after about 1840.

In between the scientific literature and the patent systems there exists a vast range of company and personal reports and letters. These are occasionally helpful but are largely ephemeral and even in the 20th century played only a minor role in documenting technology.[803]

With these caveats in mind, an examination of the 150 individuals who made major contributions to the transport changes in the 19th century indicated a number of interesting factors. In terms of birth year they were spread fairly uniformly between 1760 and 1860. The 1760 is understandable, but the 1860 date suggests that there were relatively few young contributors towards the end of the century. Perhaps the change process had run out of momentum? The youngest person in the list is Louis Renault who began designing and building innovative cars in about 1898 when only 21 years of age (Chapter 16b). Apart from being technically precocious, he was described in a 1956 issue of Time Magazine as "*vulgar, loud, domineering, impatient*".[804] It is not surprising to find precocious talent amongst people who made such major contributions. Indeed, over a quarter were under 30 when they first made a major mark on 19th century transport.

I opened this Chapter quoting Carlyle's "*great men*" of history hypothesis and I could certainly argue that many of the contributors discussed in this book could be considered as "*great men*." It would seem that some would take a contrary view, particularly of the role of science and technology. For example, there are 59 Englishmen in my list but Kitson Clark in his book on the making of Victorian England mentions only one of them – George Stephenson – in his 300 page review.[805]

The birth-place of the 150 contributors matches the story told in the preceding Chapters. 37% were born in England and 13% in Scotland, making a neat 50% from Britain. 23% were born in France and 11% in Germany. The lower German contribution reflects the fact, apparent from the preceding Chapters, that technological change did not take hold there until the second half of the century. The remaining major contributor at 11% was the USA, as it gradually emerged as a great nation. Not surprisingly, the two world cities of London (10%) and Paris (8%) were the birthplaces of many contributors. One region which stood out (7%) was the eastern portion of France, bordering on Switzerland and Germany. However, the subsequent French development was strongly focused on Paris.

In terms of where the most innovations occurred (by a crude measure) the two world cities of London and Paris dominated with each hosting 20%. The next two productive areas were in Germany, each at 5%: The home of the internal-combustion car in south western Germany and the strong academic clusters in the Berlin region. By nation, Britain contributed 47%, France 24%, and the USA and Germany both 11%. These figures are consistent with the story told in the text. Using the periods during which the 150 individuals had significant impacts, it is possible to estimate when change was occurring. In terms of periods of change, the first 20 years of the 19th century were relatively steady, as the realities of steam power became increasingly apparent. Change in the last eighty years was far more dramatic with a slight slow-down in the 1840s, possibly as steam power was more soberly digested by the market place.

PART 19

THE CONTEXT OF CHANGE

CHAPTER 19A

UNDERSTANDING CHANGE

A common engineering approach to history is a causative one: that is, after event A happened, it subsequently caused event B, which in turn later caused event C. Oh, and there might have been some humans involved. My adherence as an engineer to this cause and effect model is softened in two key ways.

First, it is important to understand why event A happened at a particular time, in a particular place, and involved a particular group of people. And why do we so often find that in retrospect event A, although never seen before, was almost inevitable at a certain time and in a certain set of circumstances?

Second, when event A does happen, often a whole suite of consequential events subsequently occur. At the time there was no obvious path to B, the next major event to retrospectively catch our eye. The cause and effect chain that seems so obvious in such a retrospective perspective, was far from obvious to those actually involved in event A. Indeed, to many at the time, event A was seen as a final culminating step, and our event B was of subsidiary and incidental importance.

Another way of putting this is that, although event B could only have happened if event A had previously occurred, event A might merely have facilitated event B and not played any role in causing it.[806] Whilst this is a plausible circumstance, it is difficult to envisage a real situation where a knowledge of event A has not at least been a spur to the development of event B.

A "*chain reaction*" occurs when each new event then creates its own new event. The above cause and effect model with event A causing event B implies a chain reaction system with a steady rate of change, which might loosely be described as linear. In terms of the 19th century transport market "*as long as business and consumer expectations were high, one technological change generated another*".[807] However, one negative response can eliminate previous gains. By definition, a successful application will be part of, and possibly the end of, a chain reaction because many new ideas will wither on the vine for a variety of reasons – particularly the lack of a marketable application for the idea.

A number of interesting facets of the chain reaction process occurred in Britain late in the 1830s and 1840s. The first is an overall effect observed by Kitson Clark[808] who noted the chain reactions that had occurred and observed that the Industrial Revolution was not marked by individual events such as the development of the steam engine at one extreme or of world markets at the other but "*to the moment when all these things combined ... to produce a new community with possibilities and realities different from anything that had existed before.*"

The second facet is that the processes were often incestuous, as noted in Chapter 8. For example, the search for useful metallic ores had required increasingly deep mines. These required more powerful pumps to continually remove water from the mine-shafts. Pumps driven by steam engines were the solution, but to make them required iron and to operate them required coal. To make iron also required coal. To move the coal and the iron required railways and the railways required iron and coal for their creation and operation. At many stages, the biggest users of coal and iron were the locomotives hauling trains carrying coal from mine to market. Of course investors could take their cut from each stage of each process and ordinary folk now had new work to do.

A third facet relates to an over-reaction to changes. New industries created by the Industrial Revolution often produced healthy profits and there was a great deal of new capital looking for new investments. For example, as discussed in Chapter 11, proposals for new steam-powered railways caught investors' attentions and the first two projects were immediately profitable, producing annual dividends of about 10 per cent. More money was thrown at increasingly marginal projects; subsequently many of the later investments soured and the "*railway mania*" bubble burst.

A fourth facet is that when an event causes multiple new events or a family of new events; the observed change is commonly no longer *linear* and the process could loosely be called *exponential*. Rosen[809] makes this point strongly with respect to the Industrial Revolution. He noted that inventions have occurred throughout history. Their flames would burn brightly and for long periods but there was never a sense that they would set alight other fires in a tinder-dry human landscape. However, in the Industrial Revolution an inventive flame could spark a fire-front of new creative ideas and inventions although most of the important changes grew very slowly immediately after their initial spark. The wild-fire process then proceeded out of any human control, with neither intervention at one extreme nor strategic goal-setting at the other.

Nevertheless it is clear from the story of 19th century transport changes presented in this book, or from a more general review of Industrial Revolution, that – at the time scale of individual people – the changes that occurred were more gradual than sudden. Indeed, there were no physical changes that could be called cataclysmic.

Of course, the overall consequential effects of the changes occurred at an increasing rate until the processes consumed all possible opportunities for change. This process is commonplace in nature where it is represented by the S curve of the *logistic equation*. The period of relatively rapid change occurs between periods of slow creation and slow extinction. Similarly, in terms of catastrophe theory, at a macro-level there were no sudden changes as a system tipped from one physical state to another.

War can cause rapid change. The British victory at Waterloo in 1815 had an immediate impact but its greatest impact in terms of transport and technology was to allow Britain to dominate international trade and to focus on the wealth-producing processes that were a consequence of that trade. Similarly, the European conquests of Napoléon imposed French organisational and educational processes on much of continental Europe. Some of these stultified innovation but, as discussed in Chapter 7, many made it possible for continental Europe to gradually gain some technical ascendancy over Britain towards the end of the century.

CHAPTER 19B

THE INDUSTRIAL REVOLUTION

At this late stage it is useful to step back and review the role of the Industrial Revolution in the processes described earlier in the book. From most viewpoints the Industrial Revolution was the world's greatest period of peaceful man-made change. Nevertheless, there is strong ongoing debate over whether it is appropriate to use the label "Revolution" for the events that occurred, mainly because the events were too localised and specific to be termed a Revolution. This consideration does not apply to transport where the events described earlier were widespread and affected most communities. Thus, this book stays with the concept of describing transport changes as revolutionary.

It is commonly accepted that the Revolution began in England in about 1760 and concluded in about 1840. It thus precipitated and then overlapped the 19th century transport changes discussed in this book. It may be appropriate to see 1840 as the end of the Industrial Revolution, particularly in England, for this was the time when the industrial world moved from developing both steam power and processes for producing useful pieces of iron, to practical and profitable uses for steam power and iron. The scene moved from rapid development to rapid application. In particular, it moved from developing new machines, to using those machines to produce immediately useable products. It is also worth noting the coincidence that in England the end of the Revolution coincided with the beginning of the Victorian Era. Less auspiciously, England and Scotland were in a pronounced trade depression from 1836 to 1843. The depression was to cause much misery to those who had invested in the early promise of the Revolution.

It could be argued that the 1840s were just a pause for breath whilst electrical power and internal-combustion engines took up the running. England was certainly favoured by initial fruits of the Revolution, for a range of reasons discussed throughout the book. However, by 1850 the Revolution had embraced Germany, France, Belgium and the United States and by 1900, Italy, Sweden and Russia. It is properly called a revolution as it fundamentally altered the way people lived their lives and it was very much an engineering revolution. Nevertheless, the term "*Industrial Revolution*", coined in English-language writings by Arnold Toynbee Snr in 1878, inadequately encompasses many of the changes that occurred during the period.[810]

Another viewpoint was put by Aldcroft and Freeman[811] who argued that there had been a continuous process of "*industrial transition*" in place long before the Industrial Revolution, and noted the role of capital in transforming "*English economy and society between 1650 and 1750*". Whilst this book has provides some evidence to support that longer-tem view, it nevertheless has demonstrated that the quantum of change, particularly in transport, that occurred during the 19th century was an order of magnitude greater than anything that had previously occurred. Moreover, it was also far greater than the more cosmetic, incremental and peripheral changes of the 20th century, when only the example of aeroplane could challenge the 19th century relativities.

Economists have written extensively about the Industrial Revolution and often, very usefully, placed it in a broader economic context. As an example, the influential American economist Walt Rostow wrote "*After the railway take-offs of the third quarter of the 19th century with coal, iron and heavy engineering at the centre of the growth process – it is steel, the new ships, chemicals, electricity, and the products of the modern machine-tool that came to dominate the economy and sustain the overall rate of growth.*"[812] I would quibble with his timing, but his overview of the events is very appropriate.

However, it is necessary to step back to first half of the 19th century, to the period before Rostow's "*railway take-offs*". In a major earlier work Lewis Mumford[813] took the Revolution's precedents even further back than Aldcroft and Freeman, arguing that there had been at least seven centuries of "*mechanical*" development "*in the monastery, in the army and in the counting-house*" leading to the many manifestations of the seeds of the Industrial Revolution. He saw the period before 1700 as one in which change was driven by scientific advancement whereas the period from 1700 to 1900 – his *paleotechnic* stage – was a time when the emphasis was on solving problems rather than on extending scientific knowledge. Given our 20th century experience, it is difficult to now see the picture that Mumford painted in 1934. In the period up to 1700, the needs of those three great wealth accumulators and protectors – war, trade and mining – had driven much scientific advancement. In the 19th century the trio still had a major influence but there was also a quickly growing broad consumer market for useful and enjoyable products beyond the needs of daily subsistence. It seems clear that the availability of these consumer-based products went hand-in-hand with expanding scientific and technical knowledge.

Indeed, the dominance of inventors in the 19th century can be readily attributed to the wide growth of consumer markets and the distinct possibility of accumulating wealth via the process of invention. This was particularly so in Britain in the first half of the century when wealth was accumulating from international trade and when the government, rather than direct and regulate, encouraged investors who in turn encouraged inventors. In this climate, it has been argued that economic growth produced entrepreneurs and risk takers, rather than vice versa.[814]

In the first half of the 19th century the greatest contributions to the Industrial Revolution, and to the transport component under examination here, came from Britain. It had inherited much of the initiative from the preceding century and its government was stable and anxious to ensure a prosperous nation. On the other hand, the two other major players, France and Germany, were initially consumed by the ambitions of Napoléon and then by the aftermath of his overthrow. The old tyrannies and repressions lasted longer in Germany and its many independent States than they did in France. However, by the second half of the century, France and the German States were once again strong industrially and financially and were ambitious to gain a greater share of the spoils of the Industrial Revolution. In the three regions, the seeds of the Revolution had been planted by and the rewards reaped by the private sector – agriculturalists, property owners, industrialists and financiers. On the other hand, after a century of change, there is some evidence that the British had become resistant to even more change.

This book has focussed on iron and steam as the principal players in the Revolution, and the changes in both epitomise the dramatic rather than the steady. Iron had been made and used for millennia but nothing in its history could have foretold the changes that occurred in the decades after 1780. Similarly, the potential for using steam had been recognised for centuries, but the changes that steam engines made to transport on land and at sea and to industrial and rural production in just three decades after 1825 were extraordinary and far beyond any prior predictions. Clark suggests that ordinary people during this period recognised the magnitude of what was happening when he writes:

> "*The sense that something revolutionary had happened, that they were living in a new world with infinite and unrealised possibilities for good and evil was indeed very strong amongst those who lived in Britain in the second quarter of the nineteenth century.*"[815]

The Industrial Revolution is sometimes recalled for the unpleasant conditions faced by workers in the new factories or by the ranks of the unemployed, as hand processes were replaced by mechanical ones (Chapter 6). Despite these and other downsides, overall "*the Industrial Revolution was an age of improvement*".[816] For instance, studies of gross domestic product per person (GDP per capita) have shown that this key indicator of personal conditions remained unchanged over most of human history and has only risen significantly in the aftermath of the Industrial Revolution.[817] In the context of this book, transport's dramatic century of change could not have occurred without the Industrial Revolution.

CONCLUSIONS

Most of the published reviews of 19th century transport have been excellent and I have drawn on and, hopefully, properly acknowledged them in this work. However, they commonly have suffered from two main problems. First, many were written too close to the end of the 19th century to be able to properly assess its consequences. Second, writers almost inevitably see their own sub-field as the major source of cause, explanation and effect. I was often reminded of ancient Indian lore describing how expert opinion is created. Six blind men touch different parts of an elephant. Each then gives totally different views of the totality of what they have touched. John Saxe's 19th century rewriting of the outcome is:

So oft in theologic wars,
The disputants, I ween,
Rail on in utter ignorance
Of what each other mean,
And prate about an Elephant
Not one of them has seen.

The 20th century introduced an Heisenbergian twist to the story in which six blind elephants were asked to touch a man. Each elephant gave an identical report – *men are flat.*

This book presents an engineer's view of the transport changes that occurred during the 19th century and describes the causes and consequences of those changes. It also takes a long-term view as only then do underlying trends become obvious. In this context, it is appropriate to note that I was recently described by the eminent transport historian Gijs Mom as "*known for his longue durée genealogy of road construction and management*".[818]

Historians approach the tasks of describing events and detecting change-makers from a wide range of perspectives – for instance, from the viewpoints of politics, economics, religion or social conditions. The fact that I am an engineer clearly colours my views and influences my interpretations and I suspect that historians reading this work would wish that I had been better schooled in their methods of historical research. On the other hand, one of the drivers of my interest in transport history has been the way in which some otherwise very capable historians have misunderstood, often grossly, the engineering aspects of the issues they were describing. I first mentioned this in 1992 in the introductory comments to my book *Ways of the World*[819] and have given further examples in the body of this work.

Previous Chapters have described the first half of the 19th century as a period during which Britain dominated technical development. It has been argued that the prime reason for this dominance was that in the rest of Europe "*absolutist governments planned, promoted and protected industry in order to increase the economic and military power of the state*".[820] Inevitably, their planning was grossly inappropriate. However, Britain and its parliament were driven predominantly by the need to invest in and encourage activities that would increase Britain's wealth, collectively and individually. Instead of planning, there was freedom of action, freedom of trade and freedom to compete. The profit motive reigned supreme and delivered wealth beyond expectation.

As the 19th century progressed the rest of the world began to focus on the lessons from the British experience and its policy settings as outlined above. There was a dilemma for observers to resolve for, as a French visitor to the Great Exhibition of 1851 at London's Crystal Palace observed: "*An aristocratic country like England is successful at supplying the people, whereas France, a democratic country, is only good at producing goods for the aristocracy.*"[821] Whereas available capital in Britain was invested by entrepreneurs in various industrial schemes, in France it was largely used as loans to the State and for the upkeep of indulgent courts and bureaucracies. Things did begin to change and, as discussed in Chapter 17b, the focus in technical development in the second half of the century moved from Britain to south-western Germany, to France and to the USA. Much of that move was fuelled by the superior educational and training facilities now being provided in those countries.

Before putting forward my own conclusions about the changes in transport during the 19th century – which I trust will not surprise any reader of the preceding Chapters – I need to dismiss some contrary conclusions reached by others, particularly those of Aldcroft and Freeman in their 1983 review of "transport in the industrial revolution."[822] On page 4 they state that for the Industrial Revolution "*the picture of transport development is characterised not by phases of concentrated, radical change so much as by a flowering of past efforts to upgrade and improve, the final working out of a continuous process of refinement and readjustment.*" But the changes described in this book – for example the Royal Mail coach service, macadam roads, iron bridges, steam locomotives, steam-powered ocean-going ships, vehicles powered by internal-combustion (IC) – all were unexpected quantum jumps that were both concentrated and radical and almost totally unpredicted in ways that could never be described as parts of a "*continuous process of refinement*".

Nevertheless, in saying this I am not subscribing to the view of some economic historians that the 19th century was a period of a few macro-inventions which spawned a series of micro-inventions which continued the momentum created by the macro-invention.[823] In all the work described in this book, there were no macro-inventions. Instead there were many long series of painstaking and persistent individual steps. The inventors may have had a macro-view but they attained their objectives by steady increments rather than great leaps forward.

There are many examples in this text – in Chapters 9 & 10 for example – of increased demand spurring new transport development, consistent with a widely held view that "*in the industrial revolution, major technological change was preceded by the rapid development of traditional methods to a point at which demand exceeded capacity*".[824] However, in the key cases of passenger rail and the car, these inventions did not respond to a gradually growing market, rather they created markets that were totally new and far beyond the expectations of the transport innovators. On the other hand, whereas many of the inventions of preceding centuries had been by scientists and aimed at advancing scientific understanding, the inventions of the 19th century – particularly in Britain – were almost entirely driven by their potential for application in the broad community and thus for sales in the market-place.

The common technical thread that runs through all the events of the 19th century is the harnessing of the combustion process. Fire would always have fascinated humans. It was magical and dangerous but it was also very useful – for heating and cooking and defending and smelting. And over time this list continued to expand. But the astonishing transformative properties of fire continually caught the human attention – destroying solid wood, boiling water to create steam and, later, igniting minerals such as saltpetre to cause explosions. In particular, the potential power of the fleeting effects of steam and of explosive combustion caught some exceptional imaginations. Nevertheless that very short burst of power could only begin to be used for practical purposes towards the end of the 18th century by which time some communities had, for various reasons, developed methods which had the potential to capture and harness it to produce manageable rotary motion. Not only did the 19th century take advantage of these new skills, but it went on to find a myriad of innovative and useful ways to utilise the now widely available and powerful rotary motion.

Perhaps the most obvious outcome to our current minds was the motor vehicle. It was certainly brought to fruition in the 19th century but its impact had not yet been felt. That impact after 1900 has been well described by Flink in pre-www days of 1985:

> "*The automotive idea, combining a light, sprung, wheeled vehicle, a compact, efficient power unit, and hard-surfaced roads, has been far more revolutionary in its impact than any technological innovation emerging from the "Second Industrial Revolution".*[825]

One simple set of conclusions that I believe I share with many[826] is that the Industrial Revolution and its 19th century transport aftermath was the result of contributions from a relatively small number of exceptional people showing great imagination, creativity, insight, skill and persistence. It is noteworthy that governments and corporations made little or no positive and specific contribution to the extraordinary outcomes achieved by these exceptional people.

It is not surprising that this conclusion can also be applied to earlier times. However, it is interesting to observe that it also applies to the 20th century despite the growing agglomeration of teams devoted to research and development. Even in productive teams, one or two individuals would always stand out as those who had the major insights and discoveries. Paradigm shifts continue to arise from individual initiatives.

The automotive industry is a case in point. Despite enormous research and development efforts, cars at the end of the 20th century were the result of a long series of incremental improvement over the 19th century vehicles of Benz and Daimler. Not only were there no major differences, but obvious possible changes such as a move from control by drivers of limited abilities to a system with even minimal input from the vehicle itself, from the external environment and from other vehicles did not occur during the 20th century, although the necessary technology – radar – had been developed by 1950 and was becoming increasingly useful. The implementation drawbacks were mainly deciding who would provide and maintain the shared infrastructure and carry the operating risks.

I am not the first author to have been fascinated by the extent, the rapidity, the magnitude and the consequences of the transport changes that occurred during the 19th century – changes that I have described as revolutionary. Furthermore, having just observed and lived through 60 years of transport in the 20th century I would suggest that its changes have been less dramatic than those of the 19th century. The changes in the 20th century have been vast in their coverage and have often been dominated by situations where demand has exceeded supply. And they have been caused more by corporations than by individuals. However, compared with the 19th century the jolting, life-altering changes driven by inspired individuals have been far less common.

Nineteenth century governments sometimes facilitated and only marginally impeded the endeavours of inventors and innovators. Ordinary citizens shared little of the wealth but, as the century ended, were beginning to share many of the benefits. For instance, one of the human consequences of better transport was that it made it easier to reduce famine and disease by sharing food and medicines. However, by the end of the century this improved mortality was being countered by the improved methods of mass-destruction being made available to the armies of the world. Nevertheless, in Britain death rates (per person) halved over the course of the 19th century.[827]

Political stability also led to a greater emphasis on the peaceful applications of the new inventions, whereas prior to the 19th century many developments had been driven by militaristic needs and applications. The main exception to this was in the mining industry where the promise of quick wealth from more efficient and deeper mines did drive many of the 18th century developments in explosives for blasting rock, engines for pumping water and vehicles for hauling ore.

We have seen how the Industrial Revolution and its transport fruits were progressed by the creative and inventive skills of some very talented people. Their work did not occur in a vacuum – rather, they received many direct and indirect inputs from the world around them. To utilise this adequately required business acumen as well as creative skill. Creative invention in itself is normally a reward in itself, and this had certainly been the case in earlier times. However, in most cases the 19th century inventions also solved pressing needs and often provided generous rewards for the combination of inventive and business skills. In retrospect, the major short-coming of these new entrepreneurs was that they showed few signs of far-reaching visions and each took the world just one step forward, or sideways, as the case might be. As an instance, James Watt could be said to be the inventor of the practical steam engine (Chapter 8). Yet he was strongly of the view that his late 18th century engines had no role to play in providing on-board power for transport.[828]

A further example – and I suggest it is a key example – is that no one during the 19th century when self-powered vehicles travelling at least as fast as a galloping horse were being contemplated, gave the slightest hint that they envisaged how such transport might operate and, consequently, how the world might look if all citizens and cargoes could travel at speeds then beyond human experience.

Perhaps the most visionary of the inventors was Gottlieb Daimler who began his working life working on steam engines with their power produced by external combustion. It was clear to him that this was not going to be the solution to a future as-yet-undefined need. In travelling in search of a solution, he saw Lenoir's large and noisy IC engine at a Paris Exhibition and realised that some development of this very different concept might be just what he was seeking. For some decades he and colleagues worked on producing a better but radically different IC engine. He saw it powering ships and carts and even aeroplanes. Having produced a useful self-powered cart, he was then lured away by investors who had another vision of the transport future. Nevertheless, Daimler was perceptive enough to take his IC patents with him. As discussed earlier (Chapter 16a) when a new venture proved short-sighted, Daimler returned to his old firm with his patents and produced the first "*modern*" car, which was called a *Mercedes*. Over a century later the firm is one of the world's largest car-makers and the Mercedes marque is still a market leader. But this was not the market that Daimler had ever foreseen for his invention.

It is easy to be swayed by the romantic idea of a few wonderful innovators such as Watt, the Brunels, Telford, Faraday, Siemens, Daimler and Benz creating major and almost instantaneous quantum jumps in technological advancement. However, if we take the prime cases of the external combustion steam engine and the IC engine, what we have actually seen are long series of incremental set-backs and advances, dead ends and new markets, misunderstandings and occasional new understandings. For example, the first really useful IC engine was Daimler and Maybach's 1885 "*grandfather clock*" engine, some 90 years after the first trials of such an engine (Chapter 14b). The 90 years were spent trying various processes and fuels, gradually understanding the physics of the process, and developing the design and manufacturing techniques able to provide the wide range of operating and control components needed to perform the basic operating process ten times every second (compared with the horse taking just one step every second). Hardenburg observed that by the late 1860s most of the essential elements of the modern IC engine had been envisaged and often designed, but bringing them together into a practical reality would take another quarter century.[829] Even Daimler and Maybach had been working together for 15 years before they achieved a basic goal with the *grandfather* clock. And then another decade was to pass until 1894 when practical IC-powered cars were in production and on sale in the open market (Chapter 16b).

Another key feature of the changes of the 19th century is that they were physical changes which people could observe and touch and use. Most of the preceding great incremental changes had been more subjective – new systems of government, new religions, new social ethics such the anti-slavery movement, and new world views such as the Copernican Revolution. Europe in particular had been touched, for instance, by the Reformation, by the astonishing new music of Bach and Mozart and Beethoven, by navigators such as Drake and Cook, by the unveiling of the planetary system and by the adventures of Napoléon, but none of these revolutions affected all citizens with the immediacy, individual directness and permanency as did the fruits of the Industrial Revolution.

None of the people discussed in this book were seers – no matter how great their contribution to transport change might have been. Indeed, few saw past the immediate usefulness of their invention and rarely perceived its future impact. Furthermore, time softens many failings and it is interesting to look at some of the railways built through the beautiful, rolling English countryside in the Railway Mania era (Chapter 11b). They were undoubtedly great scars across the landscape. Today, almost two centuries later, they seem part of that landscape and form key components of many picture postcards sold to help tourists remember the scenic attraction of the area.

One lesson that can be drawn from the 19th century is that the future will always be unknown. This is no excuse for communities not to express their social aspirations and to say that whatever unexpected changes might happen in the future, and that they should operate within some generally accepted guidelines. No such guidelines existed to prevent the excesses of the 19th century railway era and in the 20th century there were no articulated aspirations that produced guidelines preventing the over-supply and over-use of motor vehicles.

The second key lesson from the 19th century – and a lesson confirmed by the subsequent century – flows from the observation that no one, despite their personal brilliance, had in anyway predicted how the future would unfold as a consequence of the changes that they and their peers were producing.

The future is as unpredictable now as it was in the 19th century. So surely we must carefully define the guidelines for any man-made changes and we must, in the words of the finance industry, always hedge our bets to cover the gamut of possible futures.

ENDNOTES

[1] Youngson 1967:73
[2] Tucker 1975
[3] Tucker 1975
[4] Alexander 1984
[5] Pritchards 1994
[6] Smil 2017: Box 4.11
[7] Crofts 1967:52
[8] Damase 1968:75
[9] Boumphrey 1939:102
[10] Rogers 1961:92
[11] Wosk 1992:52
[12] Lay 2008
[13] Smil 2017, Box 3.5
[14] Bronowski 1973:280
[15] Smil 2017: Box 3.4
[16] Bird 1969:4
[17] Lane 1935:Plate VI
[18] Lay 1992:37-41
[19] Pyne 1806+
[20] Lay 1992:32-3
[21] Meldahl 2007:13
[22] Chartres & Turnbull 1983:85
[23] Dyos & Aldcroft 1969:78
[24] Wilson 1986
[25] Armstrong & Bagwell 1983:145-154
[26] Sobel 1996
[27] Bagwell 1974:13
[28] Lay 1992:33
[29] Duckham 1983:123
[30] Lay 2010
[31] Ellis 1998:Chap 10
[32] Lay 1998
[33] Lay 1998
[34] Lay 1991
[35] Obadiah Elliot, born in 1763 in London, England.
[36] Fenton 1794
[37] Pyne 1806+
[38] Vale 1960:Ch2
[39] John Besant: birthdate not known, he died in 1791 in London, England.
[40] Rogers 1961:32
[41] Anderson 1932, Ch13

[42] Vale 1960
[43] Belloc 1926
[44] Edward Sherman, born in 1776 in Wantage, Oxfordshire, England.
[45] Arthur Jules Morin, born in1795 in Paris, France.
[46] Law 1850:62
[47] Bagwell 1974:44, Fig. 2
[48] Pyne 1806+
[49] Tarr 1969:308
[50] McShane 1994:49
[51] Freelove 1971
[52] Boumphrey 1939:101,118,120
[53] Bird 1969:175
[54] William Chaplin, born in 1787 in Rochester, Kent, England.
[55] Rogers 1961:89
[56] Barker 1993:145
[57] Bird 1969
[58] Bird 1969
[59] Watney 1961:18
[60] loc cit
[61] Hibbs 1968:17
[62] Freelove 1971
[63] Karl Drais, born in 1785 in Karlsruhe, SW Germany.
[64] Lessing 2003
[65] Ginzroth, 1817
[66] Belloc 1926
[67] Fabre 1962:65
[68] Lessing 2003:165 (copies of parts of an 1817 book by Johann K S Bauer, citation on p518)
[69] Kirkpatrick MacMillan, born in 1812 in Keir (100 km S of Glasgow), Scotland.
[70] Willard Sawyer, born in 1808 in Maidstone, Kent, England.
[71] Garrison 1991:143
[72] Mumford 1934:14
[73] Pierre Michaux, born in 1813 in Bar Le Duc (250 km E of Paris), France.
[74] Henry Lawson, born in 1852 in London, England.
[75] John Starley, born in 1831 in Albourne, Sussex, England
[76] George Cayley, born in 1773 in Scarborough, Yorkshire, England.
[77] Charles Goodyear, born in 1800 in New Haven, Connecticut, USA.
[78] Robert Thompson, born in 1822 in Stonehaven (190 km N of Edinburgh), Scotland.
[79] John Dunlop, born in 1840 in Dreghorn, (50 km SW of Glasgow), Scotland.
[80] Hill 1980
[81] Plowden 1971:24
[82] Lay 1992:143
[83] Tucker 1975
[84] Plowden 1971:24

[85] Parsons 2007
[86] The material in this Chapter is based on a more detailed account in Lay 1992, Ch3&4.
[87] Lay 1992:59
[88] Albert 1983:38
[89] Boumphrey 1939:97
[90] Lay 1992:Ch3&4
[91] Young 1771
[92] Copeland 1968:11
[93] Albert 1983:39-41
[94] Guldi 2012
[95] Albert 1983:36
[96] Arsène Jules Dupuit, born in 1804 in Fossano, NW Italy.
[97] Dupuit 1844
[98] Kurrer 2008
[99] Watkins 1891
[100] De Villefosse 1975:143
[101] Nicolas Céard, born in1745 in Bar-sur-Aube (240 km SE of Paris), France.
[102] Davies 2013, Chap 7
[103] Boelcke 1995
[104] Lamm et al 1999:5-29
[105] Schreiber 1961:168
[106] Lay 1992:Chs1&4
[107] Albert Gallatin, born in 1761 in Geneva, Switzerland.
[108] Searight 1894:Chs1&2
[109] Schreiber 1961:210-2
[110] Lay 2010, Ch6.
[111] Thomas Telford, born in 1757 in Langholm (120 km S of Edinburgh), Scotland.
[112] Smiles 1874:101
[113] Smiles 1876
[114] Anon 2007
[115] Rolt 1958
[116] Smiles 1874
[117] Telford 1838
[118] Paxton & Shipway 2007b:23
[119] Paxton & Shipway 2007a:42
[120] Rolt 1978
[121] John L McAdam, born in 1756 in Ayr (60 km S of Glasgow), Scotland.
[122] Reader 1980
[123] Reader 1980:26-9
[124] Reader 1980:31
[125] McAdam 1816+
[126] James McAdam, born in 1786 in Ayr (60 km S of Glasgow), Scotland.
[127] Reader 1980:158+
[128] Reader 1980, p158

[129] Lay 1992, pp78-84
[130] Lay 1992, pp75-83
[131] Reader 1980:161
[132] Bird 1969:56
[133] Lay 1992, Ch11
[134] Bramley 1895
[135] Parnell 1838:28
[136] Lay 1997, 52-6
[137] Lay 1992, 226-7
[138] Lay 1992, pp203-7
[139] Lay 1992:Ch7
[140] Plumb 1950
[141] Lay 1992:115
[142] Anderson 1932
[143] Tarr 1969:279
[144] Dyos & Aldcroft 1969:79
[145] Bagwell 1974:41
[146] Bagwell 1974:50
[147] Telford 1838
[148] Cragg 1997
[149] Lay 1992:88
[150] Bodley 1971:50
[151] Penfold 1980:44-6
[152] Penfold 1980:126-8
[153] Penfold 1980:58
[154] Raymond 1984
[155] Lay 1982a
[156] Hopkins 1970:120
[157] van der Merw & Avery 1982
[158] Raymond 1984
[159] Lay 1992:263
[160] Abraham Darby, born in 1678 in Woodsettle, Staffordshire, England.
[161] Hopkins 1970:176
[162] Henry Cort, born in 1741, probably in Lancashire, England.
[163] Pannell 1964:277-285
[164] James Neilson, born in 1792 in Shettleston, Glasgow, Scotland.
[165] Biggs 1987:221
[166] More 2000:105
[167] John Wilkinson, born in 1728 in Little Clifton, Cumbria, England.
[168] Biggs 1983:219
[169] Henry Maudslay, born in 1771 in Woolwich, London, England.
[170] Ransom 1984:125
[171] Robert Mushet, born in 1811 in Coleford, Gloucestershire, England.
[172] Robert Hadfield, born in 1858 in Sheffield, England.
[173] Abraham Darby, born in 1750 in Coalbrookdale, Shropshire, England.

[174] Hopkins 1970:80
[175] Trinder 1979
[176] Billington 1983:31-2,39
[177] Smiles 1867
[178] Billington 1981, Smiles 1874:203
[179] Cox & Gould 1998:44-5
[180] James Finley, born in 1756 in Ireland.
[181] Samuel Brown, born in 1776 in London, England. This is not the Samuel Brown who worked on IC engines (#600).
[182] Photo © Philip Halling (cc-by-sa/2.0)
[183] Hopkins 1970:187
[184] Timoshenko 1953:73
[185] Smiles 1867
[186] Steinman & Watson 1957:138
[187] Louis Vicat, born in 1786 in Nevers (250 km S of Paris), France.
[188] Lay 1992:289
[189] Marc Seguin, born in 1786 in Annonay (540 km SE of Paris), France. He was a nephew of Montgolfier, the first balloonist.
[190] Guillaume-Henri Dufour, born in 1787 in Konstanz, SW Germany (at the time it was part of Switzerland).
[191] Joseph Chaley, born in 1795 in Ceyzérieu (510 km SE of Paris), France.
[192] Raymond 1984:184
[193] Henry Bessemer, born in 1813 in Hitchin, Hertfordshire, England.
[194] Raminez 2016:14
[195] Bagwell 1974:116
[196] Carl Siemens, born in 1823 in Gehrden in NW Germany. He was a cousin of Werner Siemens.
[197] Percy Gilchrist, born in 1855 in Lyme Regis, Dorset, England.
[198] Sidney Thomas, born in 1850 in London, England.
[199] Rostow 1990:61
[200] Raymond 1984:185
[201] Robert Stephenson, born in 1793 in Willington Quay, Durham (Tyneside), England.
[202] Ryall 1999
[203] William Fairbairn, born in 1789 in Kelso (70 km SE of Edinburgh), Scotland.
[204] Eaton Hodgkinson, born in in 1789 in Anderton, Cheshire, England.
[205] Timoshenko 1953:157
[206] Petroski 1994
[207] Petroski 2007, Akesson 2008
[208] Akesson 2009
[209] Petroski 1994
[210] Lay 1982a, Chap 4
[211] Euler 1744
[212] Leonhard Euler, born in 1707 in Basel, N Switzerland.
[213] Kurrer 2008

[214] Lay 1991
[215] Thomas Tredgold, born in 1788 in Brandon, Durham, England.
[216] Tredgold 1824
[217] Timoshenko 1953:266-83
[218] Henry Clifton Sorby, born in 1828 in Sheffield, England.
[219] David Kirkaldy, born in 1820 in Dundee (100 km N of Edinburgh), Scotland.
[220] August Wöhler, born in 1819 in Soltau (80 km S of Hamburg), Germany.
[221] Timoshenko 1953:171
[222] Johann Bauschinger, born in 1833 in Nuremburg, SE Germany.
[223] Friedrich Engesser, born in 1848 in Weinheim (near Mannheim, Fig. 85) in Germany.
[224] Petroski 2007
[225] Macquorn Rankine, born in 1820 in Edinburgh, Scotland.
[226] F. Braithwaite 1854
[227] Timoshenko 1953:16-7
[228] Kuzmanovic 1977
[229] Martin & Macleod 1995
[230] Brown 2003
[231] Akesson 2008:Ch2
[232] Lay 1982a
[233] Schivelbusch 1979:125
[234] Timoshenko 1953:165
[235] James B Eads, born in 1820 in Lawrenceburg, Indiana, USA.
[236] John Fowler, born in 1817 in Wadsley, Yorkshire, England.
[237] Benjamin Baker, born in 1840 in Frome, Somerset, England.
[238] Steinman and Watson 1957:252
[239] Hopkins 1970, Chap 4
[240] Lay 1992:292
[241] Marc Brunel, born in 1769 in Hacqueville (80 km NW of Paris), France. A beleaguered Royalist, Brunel immigrated to America in 1792 and became a US citizen. He moved to Britain in 1799.
[242] Joseph Monier, born in 1823 in Saint-Quentin-la-Poterie (40 km W of Avignon), France
[243] Thaddeus Hyatt, born in 1816 in Rahway, New Jersey, USA.
[244] Peters 1996:66-71
[245] Kurrerr 2008:499
[246] Gustave Wayss, born in 1851 in Mülhausen an der Enz (50 km E of Karlsruhe), Germany.
[247] Peters 1996:77
[248] François Hennebique, born in 1842 in Neuville-Saint-Vaast (50 km S of Lille), France.
[249] Mumford 1934:152
[250] Young 1785
[251] Wosc 1992
[252] Burritt 1868, p

[253] Edmund Cartwright, born in 1743 in Marnham, Nottinghamshire, England.
[254] Erasmus Darwin, born in 1731 in Newark-on-Trent, Nottinghamshire, England. He was Charles Darwin's grandfather.
[255] Wosc 1992:7
[256] James Nasmyth, born in 1808 in Edinburgh, Scotland.
[257] Nasmyth 1897:175
[258] Timoshenko 1953:54-5
[259] Charles-Louis Navier, born in 1785 in Dijon (320 km SE of Paris), France.
[260] Navier 1823
[261] Timoshenko 1953:86
[262] Euler 1744
[263] Kurrer 2008:33
[264] Forbes & Mahon 2014: 33
[265] Fairbairn 1840
[266] Hodgkinson 1840
[267] Peters 1996:61
[268] Rankine 1908
[269] e. g, Harper 2016:93 for masonry dams.
[270] Tredgold 1824
[271] Lewis Gordon, born in Edinburgh in 1815
[272] Timoshenko 1953:209
[273] Cowan 1993
[274] Lay 1982b:Section 6
[275] Lay 1982b:Section 6
[276] Timoshenko 1953:296
[277] Vitruvius c50BC, Heyman 1980
[278] Billington 1983:43
[279] Gustave Eiffel, born in 1832 in Dijon (320 km SE of Paris), France.
[280] Burnet 1974:304-12
[281] Timoshenko 1953:130
[282] Picon 2004:422
[283] Bronowski (1973:262)
[284] Mumford 1934:215
[285] Charles Coulomb, born in 1736 in Angoulême (450 km SW of Paris), France.
[286] Peters 1996:37
[287] Timoshenko 1953:43-5
[288] Humber 1857
[289] Watkins 1891:31
[290] Steinman & Watson 1957:168
[291] John Roebling, born in 1806 in Mühlhausen/Thuringen, central Germany. He moved to USA in 1831.
[292] Rosen 2010
[293] Bernstein 2004:7
[294] Timoshenko 1953:126
[295] Perkin 1969:412

[296] Kurrer 2008:50
[297] Rostow 1990:61
[298] Singer et al 1958, vii
[299] More 2000:170
[300] Benjamin Thompson, born in 1753 in Woburn, Massachusetts, USA. A Royalist, he moved to England in 1783 after the American War of Independence. He moved to Bavaria in 1785and between 1799 and his death in 1814 he divided his time between England and France. Whilst in Bavaria he was made Count Rumford.
[301] James Joule, born in 1818 in Salford, Lancashire, England.
[302] William Thompson, born in 1824 in Belfast, Northern Ireland. He was made Lord Kelvin in 1892.
[303] Nicolas Carnot, born in 1796 in Paris, France.
[304] Garrison 1991:147
[305] Rudolf Clausius, born in 1822 in Koszalin (near the Baltic coast), Poland.
[306] Carnot 1824: 38&42
[307] Bruckberger 1959:181-2
[308] Ransom 1984
[309] Benoit Fourneyron, born in 1892 in Pont-sur-l'Ognon (380 km SW of Paris), France.
[310] Garrison 1991:143
[311] Charles Parsons, born in 1854 in London, England.
[312] Hardenberg 1993:14-26
[313] Christiaan Huygens, born in 1629 in The Hague, Holland.
[314] Hardenburg 1993:37-53
[315] Denis Papin, born in 1647 in Chitenay (190 km S of Paris), France.
[316] Hardenburg 1993:124-25
[317] Thomas Savery, born in 1650 in Modbury in Devon, England.
[318] Black and Davis (1913).
[319] Thomas Newcomen, born in 1664 in Dartmouth, Devon, England.
[320] Schivelbusch 1979:2
[321] No bio data available
[322] Nicholas Cugnot, born in 1725 in Void-Vacon (270 km E of Paris), France.
[323] de Villefosse:267
[324] James Watt, born in 1736 in Greenock (near Glasgow), Scotland.
[325] Matthew Boulton, born in 1728 in Birmingham, England.
[326] Clarke 1965:39
[327] Thompson1963:20
[328] Uglow 2002:257
[329] Oliver Evans, born in 1755 in Newport, Delaware, USA.
[330] Ransom 1984:19 & Fletcher 1891:4-7
[331] Lay 2009, Ch11
[332] Clarke 1965:34 & Fletcher 1891:16
[333] Carnot 1824:39
[334] Schivelbusch 1979:5-6

[335] Kuznets (1930:2+)
[336] William Murdock (or Murdoch), born in 1754 in Cumnock (60 km S of Glasgow), Scotland.
[337] Richard Trevithick, born in 1771 in Tregajorran, Cornwall, England.
[338] Everett 1955:19
[339] Beasley 1988:6
[340] Lay 1992:35
[341] Lewis 1970
[342] Lewis 1970
[343] Ransom 1964:62
[344] Ransom 1990:13
[345] Lewis 1970, Plate 51
[346] Ransom 1990:15
[347] Ransom 1984:58
[348] John Birkinshaw, born c1780 in Bedlington, Northumberland, England
[349] Ransom 1984:53
[350] Williams 1883:236
[351] Kellys 2006:159
[352] Robert Stevens, born in Hoboken, New Jersey, USA in 1787. He is the son of John Stevens (endnote 601)
[353] Ransom 1990:225
[354] Lewis 1970
[355] Lay 2009, Chs8&11
[356] Lay 2009, Ch13
[357] Alphonse Loubat, born in 1799 in Saint-Livrade-sur-Lot (600 km S of Paris), France.
[358] John Blenkinsop, born in 1783 in Felling, County Durham, England.
[359] Matthew Murray, born in 1765 in Newcastle.
[360] William Hedly, born in 1779 in Newburn near Newcastle upon Tyne, England.
[361] Ransom 1984:128
[362] Charles Brandling, born in 1733 in Gosforth, Northumberland, England.
[363] Ransom 1984:173
[364] Richard Reynolds, born in 1735 in Bristol, England.
[365] John Curr, born in 1756 in County Durham, England.
[366] Ransom 1984:52
[367] Reader 1980:169
[368] Williams 1885:6-7
[369] McShane 1994:11
[370] Beasley 1988:6
[371] e. g. Williams 1885:32
[372] Williams 1885:330
[373] Dickinson & Tilley 1934:108; Ransom 1984:84
[374] Bagwell 1974:90
[375] Bagwell 1974:91
[376] George Stephenson, born in 1781 in Wylam, Northumberland, England

[377] Steinman & Watson 1957:147
[378] Simmons 1975:3
[379] Williams 1885:11
[380] Schivelbusch 1979:4
[381] Simmons 1975:49
[382] Carnot 1825:40
[383] Walter Hancock, born in 1799 in Marlborough, Wiltshire, England.
[384] Goldsworthy Gurney, born in 1793 in Padstow, Cornwall, England.
[385] McCausland 1948:105
[386] Dalgleish 1980:122
[387] Beasley 1988:143
[388] Francis Maceroni, born in 1788 in Manchester, England.
[389] Fletcher 1891:120-8, Anderson 1932, Maceroni 1835
[390] Nathaniel Ogle, born in 1796 in Southampton, England.
[391] Copeland 1968
[392] Everett 1955:24
[393] William James, born in 1796 in Henley, near London, England.
[394] Bird 1960:18: see also Fletcher 1891:74-82
[395] Barker & Robbins 1963:45
[396] Fletcher 1891
[397] Bird 1969:173
[398] John Grimshaw, born in 1763 in Sunderland, County Durham, England.
[399] Ransom 1984:73-9
[400] John Fitch, born in 1743 in Windsor, Connecticut.
[401] William Symington, born in 1764 in Leadhills, South Lanarkshire (70 km SW of Edinburgh), Scotland.
[402] Robert Fulton, born in 1765 in Lancaster County, eastern Pennsylvania, USA.
[403] Ransom 1984:90
[404] Henry Bell, born in 1767 in Torphichen (between Edinburgh and Glasgow), Scotland.
[405] Ransom 1984:121-4
[406] Barker & Robbins 1963:41
[407] David Napier, born in 1790 in Camlachie, Glasgow, Scotland.
[408] Pettit Smith, born in 1808 in Hythe, Kent, England.
[409] Ransom 1984:179
[410] Armstrong & Bagwell 1983:171
[411] Isambard Kingdom Brunel, born 1806 in Portsmouth, Hampshire.
[412] Kellys 2006
[413] William Froude, born in 1810 in Darlington, Devon, England.
[414] Dyos & Aldcroft 1969:248
[415] Pannell 1954:144-156
[416] Jackson 1981
[417] Arthur Woolf, born in 1766 in Camborne, Cornwall, England.
[418] John Elder, born in 1824 in Glasgow, Scotland.
[419] Dyos & Aldcroft 1969:259

[420] Dyos & Aldcroft 1969:262
[421] Melosi 1982
[422] Dyos & Aldcroft 1969;122
[423] Garfield 2002:14
[424] Crofts 1967:3
[425] Bagwell 1974:59
[426] Williams 1885:15 & Bagwell 1974:91
[427] Beasly 1988:43
[428] Bagwell 974:96
[429] Lay 1992:34
[430] Ransom 1990:46
[431] Williams 1885:17
[432] Ransom 1990:52
[433] Williams 1885:275 & Ransom 1990:402
[434] Lay 2004:22
[435] Lay 2009:Ch19
[436] Lay 2003:176
[437] Wilson 1885:284&286
[438] Lay 2009, Ch16
[439] Garfield 2002
[440] Simmons 1975:i
[441] Labrum 1994:217-9
[442] Cragg 1997:15-7
[443] Rogers 1961:193
[444] Beasley 1988:39
[445] William Otis, born in 1813 in Pelham, Massachusetts, USA.
[446] Grimshaw 1985:11-12
[447] Beasley 1988:40
[448] Neuberg 1977
[449] Briggs 1987:248
[450] Simmons 1975:54
[451] Simmons 1975:51
[452] Beasley 1988:39
[453] Bagwell 1974:4-6,163
[454] Beasley 1988:39
[455] Dalgleish 1980:117
[456] Williams 1885:41,55
[457] Dyos & Aldcroft 1969:127
[458] George Hudson, born in 1800 in Howsham, Yorkshire, England.
[459] Williams 1883:48
[460] Reed 2004
[461] Schivelbusch 1979:28
[462] Witcher 2017
[463] Armstrong 1976:132-3
[464] Gordon 2016:Chapter 5

[465] Schivelbusch 1997:xiii
[466] Daniels & Rose 1982
[467] Cipolla 1978:79
[468] Bible RSV, Genesis 24:60
[469] Syme 1952
[470] Illustrated London News, 11 Feb 1843:93
[471] Lay 2009:68
[472] Rogers 1961:93
[473] Lay 2003:43
[474] Schreiber 1961:179
[475] Pettifer & Turner 1984:99
[476] Bagwell 1974:143
[477] Inwood 2005:225
[478] McShane 1994:51
[479] McShane 1994:44
[480] Bagwell 1974
[481] McShane 1994:25
[482] Beasley 1988:129
[483] Dyos & Aldcroft 1969:128-57
[484] Thompson 1963:256
[485] Bradshaw 1863:II.2
[486] Mom 2007:35
[487] Beasly 1988, Appendix
[488] Anderson 1932
[489] At the end of Chapter 7. The book was first published by Punch magazine during 1847-48.
[490] Bagwell 1974:107
[491] Chartres & Turnbull 1983:66
[492] George Pullman, born in 1831 in Brocton, New York State, USA.
[493] Rogers 1961:115,7&9
[494] Wosk 1992:58
[495] Williams 1885:387
[496] Williams 1885:372
[497] Dyos & Aldcroft 1969:Chap 6
[498] Marx 1857. Grundrisse, p538
[499] Rostow 1990:62
[500] Rogers 1961:77
[501] Georgano1972:94
[502] Ransom 1990:174
[503] Williams 1885:334
[504] Rogers 1961:122
[505] Ransom 1990:190
[506] Ellis 1958:Ch15
[507] Ransom 1990: 3
[508] Georgano 1972:97

mford 1934:162
Wosk 1992:67
[511] Sobel 1996
[512] Simmons 1975:81
[513] George Westinghouse, born in 1846 in Schoharie, New York State, USA.
[514] Rogers 1961:180
[515] Rogers 1961:202
[516] Nock 1972
[517] Sekon 1938:5,134
[518] Dickens 1848:Ch6, opening paragraphs
[519] Bradshaw 1863:I:5
[520] Sekon 1938:6
[521] Barker & Robbins 1963:xxviii
[522] Barker and Robbins 1963:57+
[523] Law 1850:3
[524] Railway Times, 29 June 1850 in Barker & Robbins 1963:54
[525] Barker & Robbins 1963:47
[526] Smith 2001:152
[527] Bagwell 1974:133
[528] Barker & Robbins 1999:241
[529] Inwood 2005:2
[530] Sekon 1938:8,9
[531] John Rennie, born in 1761 in Phantassie, Scotland.
[532] Lay 2009, Ch20
[533] Barker & Robbins 1963:66
[534] Doré 1872
[535] Inwood 2005:2
[536] Doré 1872
[537] Lay 1992: 129
[538] George Shillibeer, born in 1797 in London, England.
[539] Hibbs 1968
[540] Dunbar 1961:21
[541] Richard Stephens, born in 1856 in Cwmbran, Wales.
[542] Joseph Hansom, born in 1802 in York, England.
[543] Dickens, C, 1836. Hackney coach stands
[544] Moore 1902: 217
[545] Thompson, J. and Smith, A. 1877
[546] Buchanan 1958:109, Whitfield 2006:153
[547] Edwards 1898
[548] Ackroyd 2000:517
[549] Reader 1980:167
[550] Whitfield 2006:99
[551] Hussey 2006:262
[552] Georges-Eugène (later Baron) Haussman, born in 1809 in Paris, France.
[553] Hussey 2006:266,

[554] Dunbar 1967
[555] John Stephenson, born in 1803 in County Armagh, Ireland.
[556] Dunbar 1967:16
[557] Bird 1969:135
[558] McShane 1994:14
[559] McShane 1994:1
[560] McShane 1994:12
[561] Dunbar 1967
[562] George F Train born in 1829 in Boston, USA.
[563] Sekon 1938, op p594
[564] Hibbs 1968:11
[565] Inwood 2005:225
[566] Andrew Hallidie, born in 1836 in London, England, England.
[567] Werner von Siemens, born in 1816 near Hannover, NW Germany. He was a cousin of Carl Siemens.
[568] Frank Sprague, born in 1857 in Milford, Connecticut, USA.
[569] Hilton 1969
[570] McShane 1994:15
[571] Barker & Robbins 1963:45
[572] Barker & Robbins 1963:99
[573] Beasley 1988:43
[574] Hibbs 1968:21
[575] Beasley 1988:47
[576] Penfold 1979:119
[577] Petroski 1996
[578] Hewison 1983
[579] Stein 1973:33
[580] Beasley 1988:57
[581] Bird 1969:172
[582] For example, see Maceroni 1835
[583] Fletcher 1891:2
[584] Mountfield 1976:167
[585] Buchanan 1958:2
[586] All that is known of Thomas Rickett is that he worked at the Castle Foundry, Buckingham, England in the 1850s and 1860s.
[587] Booth 1971:34
[588] Beasley1988:55
[589] Plowden 1971:21
[590] Lay 2009:§27.7.3
[591] Lay 2009:§19.2
[592] Lay 2009:§19.2
[593] Lay 2009:§27.7
[594] Lay 2009:§19.2
[595] Lay 2009:§19.2
[596] McShane 1994:11

[597] Georg Lankensperger, born in 1799 in Marktl, Oberbayern, SE Germany
[598] Georgano 1972:25
[599] Thomas Aveling, born in 1824 in Elm, Cambridgeshire, England.
[600] Lay 1992:209
[601] Antoine-Rémy Polonceau, born in 1778 in Reims (150 km NE of Paris), France.
[602] E.g. McShane1994:Ch5
[603] Aveling-Barford 1965
[604] Dunbar 1967
[605] McShane 1994:97
[606] Amédée Bollée, Snr, born in 1844 in Le Mans (210 km SW of Paris), France.
[607] Léon Serpollet, born in 1848 in Culoz (530 km SE of Paris), France.
[608] Jules-Albert de Dion, born in 1856 in Carqufou (370 km SW of Paris), France.
[609] Bird 1960:44
[610] George Bouton, born in 1847 in Léon (730 km SW of Paris), France.
[611] Charles-Armand Trépardoux, born in 1853 in Paris, France.
[612] Beasley 1988:83
[613] Bird 1960:47
[614] Eckermann 2001:18
[615] Eckermann 2001:18
[616] Twins Francis and Freelan Stanley, born in 1849 in Kingfield, Maine, USA.
[617] Foster 2004, xxii & 54
[618] Robert Street. No bio data found. Possibly born about 1760 in London, England.
[619] John Stevens, born in 1749 in Hoboken, New Jersey, USA. His son Robert was also an inventor (see endnote 344)
[620] Hardenberg 1999:Chaps 2 & 3
[621] Phillippe Lebon d'Humersin, born in 1767 in Brachay (270 km E of Paris), France. Sometimes recorded as Humbersin.
[622] Isaac de Rivaz, born in 1752 in Paris, France.
[623] Hardenberg 1999:Ch8
[624] No data is available on Samuel Brown's birth date or location (probably London, England). This is not the same Samuel Brown who worked on iron chains (#177).
[625] Hardenberg 1999:Ch12
[626] Hardenberg 1999:77
[627] Samuel Morey, born in 1762 in Hebron, Connecticut, USA.
[628] Hardenberg 1999:Ch13
[629] There is some doubt about Lemuel Wright's bio. He was working in London, England between 1820 and 1860, but appears to have been born in Plainfield, New Hampshire, USA in 1790 and died in Brooklyn, NY, in 1886.
[630] Hardenberg 1999:Ch16
[631] William Barnett, born in 1802 in Bradford, England.
[632] Eugenio Barsanti, born in 1821 in Pietrasanta (near Pisa), Italy.
[633] Felice Matteucci, born in 1808 in Lucca (near Pisa), Italy

[634] Hardenburg 1999:Ch24
[635] Etienne Lenoir, born in 1822 in Mussy-la-Ville in Luxembourg.
[636] Hardenberg 1999:235
[637] Hardenberg 1999:249
[638] Bourdelin 1860:395
[639] Anon 1860
[640] Hardenberg 1999:255
[641] Bourdelin 1860:395
[642] Hardenburg 1999:229
[643] Nikolaus Otto, born in 1832 in Holzhausen an der Haide, Germany (see Fig. 85).
[644] Eckermann 2001:20
[645] Alphonse-Eugène Beau de Rochas, born in 1815 in Digne-les-Bains (140 km NE of Marseille), France.
[646] Hardenburg 1999:Ch28
[647] Eugene Langen, born in 1833 in Cologne, Germany.
[648] Hardenberg 1999:Ch33
[649] Gottlieb Daimler, born in 1834 in Schorndorf (near Stuttgart) in Germany
[650] Flower & Jones 1981:13
[651] Simsa 1986:10-15
[652] Wilhelm Maybach, born in 1846 in Heilbronn (near Stuttgart) in Germany.
[653] Eckermann 2001:22
[654] Hardenberg 1999: Chs34-7
[655] Christian Reithmann, born in 1818 in St Jakob in Haus (70 km SE of Salzburg), Austria.
[656] Hardenberg 1999:286
[657] Hardenberg 1999:468
[658] Hardenberg 1999:471
[659] Eckermann 2001:279-283
[660] Georgano 1972:62
[661] Eckermann 2001:29
[662] Simsa 1986:124
[663] Flower & Jones 1981:15
[664] Georgano 1972:38
[665] Georgano 1972:52
[666] Karl Benz, born in 1844 in Mühlberg, a suburb of Karlsruhe, Germany.
[667] Flower & Jones 1981:12
[668] Mercedes-Benz AG 1992:11
[669] Simsa 1986
[670] Edward Butler, born in 1862 in England (possibly in Greenwich).
[671] Bird 1969:202
[672] Barker 1993:146
[673] Simsa 1986:4
[674] Lay 1992:Ch7
[675] Aimé Argand, born in 1750 in Geneva, Switzerland.

[676] Johan Becher, born in 1635 in Speyer (15 km S of Mannheim, Fig. 85) in Germany.
[677] Hardenberg 1999:Ch10
[678] Eilhrad Mitscherlich, born in 1794 in Wilhelmshaven (a Baltic port), Germany.
[679] Michael Faraday, born in 1791 in London, England.
[680] Edwin Drake, born in 1819 in Greenville, New York, USA.
[681] Lucier 1991
[682] Julius Hock, born in 1848 in Vienna.
[683] Alessandro Volta, born in 1745 in Como, N Italy.
[684] Hans Christian Ørsted, born in 1777 in Rudkobing, Denmark.
[685] Forbes & Mahon 2014: Ch 5
[686] Forbes & Mahon 2014: 100
[687] William Sturgeon, born in 1783 in Whittington, Lancashire, England.
[688] James Clerk Maxwell, born in Edinburgh, Scotland, in 1831.
[689] Moritz Hermann Jacobi, born in 1801 in Potsdam, near Berlin in Germany.
[690] Robert Davidson, born in 1804 in Aberdeen, Scotland.
[691] John Daniell, born in1790 in London, England.
[692] Paul (or Pavel) L Schilling, born in 1786 in Tallin, Estonia. He became Baron Schilling von Cannstatt.
[693] Ransom 1990: Chapt 6
[694] Charles Wheatstone, born in 1802 in Gloucester in England.
[695] William Cooke, born in 1806, in Durham, England.
[696] Samuel Varley, born in 1832 in London, England.
[697] Sebastian de Ferranti, born in 1864 in Liverpool in England.
[698] Smil 2005:83
[699] Raymond Planté, born in 1834 in Orthez (800 km S of Paris), France.
[700] Nicolas Raffard, born in 1824 in Paris, France.
[701] Thomas Parker, born in 1843 in Coalbrookdale, Shropshire, England.
[702] Alexander Graham Bell, born in 1847 in Edinburgh, Scotland. He emigrated to Canada after completing his academic studies.
[703] Thomas Edison, born in 1847 in Milan, Ohio, USA.
[704] Guglielmo Marconi, born in 1847 in Bologna, N Italy.
[705] Richard Roberts, born in 1789 in Llanymynech in Wales.
[706] Émile Roger, born in 1850 in Paris, France.
[707] Barker 1993:146
[708] Watney 1961:p67
[709] Bird 1960: 36
[710] Édouard Michelin, born in 1859 in Clermont-Ferrand (420 km S of Paris), France
[711] André Michelin, born in 1853 in Paris, France.
[712] Simsa 1986:19
[713] Beasley 1988, 96-100
[714] René Panhard, born in 1841 in La Bourboule (500 km S of Paris), France.
[715] Émile Levassor, born in 1843 in Marolles-en-Hurepoix (40 km S of Paris) in France.

[716] Armand Peugeot, born in 1841 in La Bourboule (460 km S of Paris), France.
[717] Louis Renault, born in 1877 in Paris, France.
[718] Beaumont (1900:87)
[719] Frederick Simms, born in 1863 in Hamburg, northern Germany.
[720] The Victoria Wagonette had been introduced into England in the 1850s by Prince Albert and had been based on an existing German carriage design.
[721] Eckermann 2001:35
[722] Rudolf Diesel born in 1858 in Paris, France but educated in Munich, Germany.
[723] Léon was Amédée Bolleé's younger brother.
[724] Eckermann 2001:39
[725] Beasley 1988:129
[726] McShane 1994:98
[727] Emil Jellinek, born in 1853 in Leipzig, eastern Germany.
[728] Condradt 1986
[729] Bird 1960:490
[730] Borth 1969:70
[731] Enrico Bernadi, born in 1841 in Verona, Italy.
[732] Charles & Frank Duryea, born 1861 & 1869 in Canton & Washburn, Illinois.
[733] John Henry Knight, born Farnham, Surrey in 1847.
[734] Frederick Lanchester, born in 1868 in Lewisham, London.
[735] Ruiz 1988:16-9
[736] Flink 1990:23
[737] Bagwell 1974:202
[738] Sachs 1992:14
[739] Beasley 1988:127
[740] Sachs 1992:3
[741] Baker, R. 1899. McClure's Magazine, 13(3):195-7, July
[742] Buchanan 1958:210 although the underlining is mine
[743] McShane 1994:x
[744] Lay 1992:313
[745] e.g. Lay 1992, 155-165
[746] Rostow (1990:60
[747] Plowden 1971:24
[748] Bird 1969:39
[749] Plowden 1971:23
[750] Lay 2003:40
[751] Guldi 2012:conclusions
[752] Buchanan 1958:109
[753] Lay 2007
[754] Lay 1992, 161-2
[755] Geddes 1940:26
[756] Lay 2009, Ch16
[757] Lay 2009, Ch19
[758] Lay 2012b
[759] Lay 1992:34-6

[760] Lay 2009:67
[761] Aitken 1900:305
[762] Lay 2011
[763] Lay 1992:176
[764] Lay 2010:172
[765] Lay 1992:197
[766] Lay 2012a
[767] McShane 1994:51
[768] Lay 1992:197-202
[769] Lay 2009:Ch17
[770] Doré 1872
[771] McShane 1994
[772] McShane 1994:xii
[773] e.g. see Dawson 1876
[774] Lay 2009:Chs8-12
[775] Lay 1992:ch7
[776] No bio data available. He is said to have come from Basle.
[777] Lay 1992, pp216-7
[778] Lay 1992, Ch11
[779] loc cit
[780] Joseph Aspdin, born in 1778 in Hunslet, near Leeds in Yorkshire, England.
[781] Lay 1992, Ch7
[782] Lay 2009:Ch31
[783] John Burgoyne, born in 1782 in London, England.
[784] Lay 2005
[785] Henry Carey, born in 1793 in Philadelphia, USA.
[786] See examples in Lay 2003a:53 and 2005:164
[787] Thompson 1963
[788] Lay 2003b
[789] Cipolla 1978
[790] Dyos and Aldcroft 1969
[791] Thompson 1963. 229
[792] Wise 2008
[793] Briggs 1987; 216
[794] Lay 1992:15
[795] Briggs 1987:230
[796] Bottles 1987:6
[797] Lay 1992:15
[798] Carlyle 1841
[799] Hegel 1837
[800] Spencer 1896:34
[801] Bardi 2006
[802] More 2000:95
[803] Lay 1988
[804] Wikipedia entry for Renault

[805] Clark 1962
[806] Clark 1962:85
[807] Briggs 1987:219
[808] Clark 1962:86
[809] Rosen 2010
[810] e.g. Perkin 1969:1
[811] Aldcroft and Freeman 1983:1
[812] Rostow 1990:59
[813] Mumford 1934:3
[814] Briggs 1987:219
[815] Clark 1962:86
[816] Thompson 1963, 214
[817] Rosen 2010
[818] Mom 2012
[819] Lay 1992:1
[820] Hirst 2009:142
[821] Briggs 1987:222
[822] Aldcroft and Freeman 1983
[823] e.g. More 2000:Chap 1
[824] Mountfield 1976:11
[825] Flink 1985
[826] e.g. Guldi 2012, 16
[827] Cipolla 1978, Table 14a
[828] Bird 1969:158
[829] Hardenburg 1999:229

REFERENCES

Ackroyd, P.2000. *London: the biography*. London: Chatto & Windus

Aitken, T. 1900. *Roadmaking and maintenance*. London: Charles Griffin

Akesson, B. 2008. *Understanding bridge collapses*. London: Taylor & Francis

Albert, W. 1983. *The turnpike trusts*. In Aldcroft and Freeman, 31-63

Aldcroft, D. & Freeman, M. (Eds) 1983. *Transport in the Industrial Revolution*. Manchester: Manchester UP

Alexander, R. M. 1984. Walking and running. *Am Sci* 72(4):348-54

Anderson, R. M. 1932. *The roads of England*. London: Benn

Anon 1860. A new caloric engine. Sci Am III(13):193, 22 Sep

—. 2007. Thomas Telford: 250 years of inspiration. *Proc Inst Civ Engrgs*, 160

Armstrong, E.L. (Ed). 1976. *History of public works in the United States 1776-1976*. Chicago: American Public Works Association

Armstrong, J. & Bagwell, P. S. 1983. Coastal shipping. In Aldcroft and Freeman, 142-176

Aveling-Barford, 1965. *A hundred years of road rollers: A pictorial record*. Lingfield, Surrey: Oakwood Press.

Bagwell, P. S. 1974. *The transport revolution from 1770*. London: Batsford

Bardi, J. 2006. *The calculus wars*. New York: Basic Books

Barker, T. C. 1993. Slow progress: forty years of motoring research. *J Trsp Hist* 14(2):142-165, Sept

—. & Robbins, M. 1963. *A history of London transport; passenger travel and the development of the metropolis*. London: George Allen and Unwin

Beasley, D. R. 1988. *The suppression of the automobile*. New York: Greenwood Press. Republished in 1997 as *Who really invented the automobile?* Simcoe, Ontario: Davus

Beaumont, W, 1900. *Motor vehicles and motors*. Westminster: Archibald Constable

Belloc, H. 1926. *The highway and its vehicles*. London: Sudio

Bernstein, W. J. 2004. *The birth of plenty*. New York: McGraw-Hill

Billington, D. P. 1981. Bridge design and regional aesthetics. *Proc. ASCE*, 107(ST3):473-86

Bird, A. 1960. *The motor car, 1765-1914*. London: Batsford
—. 1969. *Roads and vehicles*. London: Longmans
Black, N. & Davis, H. 1913. *Practical physics*. New York: Macmillan
Bodley, H. 1971. *Roads*. London: Batsford
Boelcke, W. A. 1995. The importance of the road in the 18th and 19th century. *Strabag Proc, 50, The road in the change of centuries*, Cologne, pp 46-9
Booth, G. A. 1971. Edinburgh experimental vehicles. In *The Omnibus*, J. Hibbs, Ed. Newton Abbot: David & Charles
Borth, C. 1969. *Mankind on the move*. Washington: Automotive Safety Foundation
Bottles, S. L. 1987. *Los Angeles and the automobile*. Berkeley: Univ of California Press
Boumphrey, G. 1939. *British roads*. London: Thomas Nelson
Bourdelin, E. 1860. La machine Lenoir. *Le Monde Illustré*, 16 Jun, 166:396-6
Bradshaw, G. 1863. *Bradshaw's descriptive railway hand-book of Great Britain and Ireland*. London:Adam (reprinted in 2012 by Old House in Oxford)
Braithwaite, F. 1854. On the fatigue and consequent fracture of metals. Minutes of the *Proc Instn Civil Engineers*, XIII:463-75
Bramley, R. R. 1805. *The road-maker's guide*. Leeds: G. Wright Jun
Briggs, A. 1987. *A social history of England*. 2nd Edn. London: Penguin
Bronowski, J. 1973. *The ascent of man*. London: BBC
Brown, J. 2003. Researchers offer new explanation of Tay Bridge disaster. *ASCE Civ Engrg*, Apr, 73:26-7
Bruckberger, R. L. 1959. *Image of America*. New York: Viking
Buchanan, C. D. 1958. *Mixed blessing: the motor in Britain*. London: Leonard Hill
Burnett, J. 1974. *Useful toil*. London: Allen Lane
Burritt, E. 1868. *Walks in the Black Country and its green border-land*. Birmingham: S. Low, Son, and Marston
Carlyle, T. 1841. *On Heroes, Hero-Worship and the Heroic in History*. London: James Fraser
Carnot, N. L. S. 1824. *Réflexions sur la puissance motrice de feu*. Paris: Bachelier (English translation by R H Thurston, published in New York by Wiley, 1897)
Chartres, J. A. & Turnbull, G. L. 1983. *Road transport*. In Aldcroft and Freeman, 64-99
Cipolla, C. M. 1978. *The economic history of world population*. 7th edn. Harmondsworth, Middlesex: Penguin

Clark, G. K. 1962. *The making of Victorian England*. Edinburgh: T and A Constable

Clarke, D. 1965. *The ingenious Mr Edgeworth*. London: Oldbourne

Condradt, D-M. 1986. How the name Mercedes and the star came about. *In aller Welt (Mercedes-Benz)*. 199(1):22-5

Copeland, J. 1968. *Roads and their traffic: 1750-1850*. Newton Abbot: David and Charles

Cowan, H. J. 1993. The scientific and political disputes which preceded the development of our present theory of structures. *Trans Multi-Disciplinary Engrg Aust*, GE17(1),June:51-8

Cox, R. C. & Gould, M. H. 1998. *Ireland: Civil engineering heritage*. London: Thomas Telford

Cragg, R. 1997. *Wales & west central England. Civil engineering heritage*. London: Thomas Telford, 2nd Edn

Crofts, J. 1967. *Packhorse, wagon and post: land carriage and communications under the Tudors and Stuarts*; London: Routledge

Damase, J. 1968. *Carriages*. London: Weidenfeld & Nicholson

Daniels, G. H. and Rose, M. H. 1982. *Energy and transport: historical perspectives on policy issues*. Beverley Hills: Sage.

Davies, N. 2013. *Vanished kingdoms*. London: Allen Lane & Penguin

Dawson, G. J. C. 1876. Street pavements. *J Liverpool Polytechnic Society*, 15-75. Available in ICE, Tracts 8vo, Vol. 270

de Villefosse, R. H. 1975. *Histoire des grandes routes des France*. Paris: Librairie Académique Perrin

Dickens, C. 1836. *Sketches by Boz*. London: John Macrone

—. 1848. *Dombey and son*. London: Bradbury and Evans

Dickinson, H. & Tilley, A. 1934. *Richard Trevithick, the engineer and the man*. Cambridge: Cambridge University Press

Doré, L. A. G., and Jerrold, B. 1872. *London: a pilgrimage*. London: Grant. Facsimile Editions by Blom of New York in 1968 and Dover in 2004

Duckham, B. F. 1983. *Canals and river navigation*. In Aldcroft and Freeman, 100-141

Dunbar, C. S. 1967. *Buses, trolleys and trams*. London: Paul Hamlyn

Dupuit, A. J. 1844. De la mesure de l'utilité des travaux publics. *Annales des ponts et chaussées*, Second series, 8.

Dyos, H. J. & Aldcroft, D. H. 1969. *British Transport: an economic survey from the seventeenth century to the twentieth*. Leicester: University of Leicester Press

Eckermann, E. 2001. *World history of the automobile*. Warrendale (Pa): Society of Automotive Engineers (updated English translation of 1989 German edition)
Edwards, P. J. 1898. *History of London street improvements 1855-1897*. London: County Council
Ellis, C. H. 1958. *The development of railway engineering*. Chapter 15 of Singer et al 1958
Ellis, P. B, 1998. *The ancient world of the Celts*. London: Constable
Everett, L. A. 1955. *The shape of the motor car*. London: Hutchinson
Fabre, M. 1962. *A history of land transportation*. Geneva: Nitsche (English edition in 1964 by Leisure Arts in London).
Fenton, W. 1794. *A treatise on carriages*. London: the author
Fletcher, W. 1891. *The history and development of steam locomotion on common roads*. London: Spon. Reprinted in 1972 as Steam on common roads, London: David and Charles
Flink, J. J. 1985. Innovation in automotive technology. *Am Sci* 73(2):151-61
Flower, A. & Johns, M. W. 1981. *100 years on the road: a social history of the car*. New York: McGraw-Hill
Forbes, N. and Mahon, B. 2014. *Faraday, Maxwell and the electromagnetic field*. New York: Prometheus
Foster, K. 2004. The Stanley Steamer: America's legendary steam car. Kingfield, Maine: Stanley Museum
Freelove, W. F. & Lang, J. 1971. *An assemblage of 19th century horses & carriages*. Farnham, Surrey: Perpetua
Gallatin, A. 1808. *Roads and Canals*. American State Papers: Miscellaneous. Vol 1:724-41. Washington: US Govt
Garfield, S. 2002. *The last journey of William Huskisson*. London: Faber & Faber
Garrison, E. 1991. *A history of engineering and technology: artful method*. Boca Raton, USA: CRC Press
Geddes, N. B. 1940. *Magic motorways*. New York: Random House
Georgano, G. N. (Editor) 1972. A history of transport. London: Dent
Gordon, R. J. 2016. The rise and fall of American growth. Princeton: Princeton University Press
Grimshaw, P. 1985. *Excavators*. Poole, Dorset: Blandford
Guldi, J. 2012. *Roads to power*: Britain invents the infrastructure state. Cambridge (Mass, USA): Harvard University Press
Hardenberg, H. O. 1993. *The antiquity of the internal-combustion engine: 1509 - 1688*. Warrendale, Penna; Society of Automobile Engineers

—. 1999. *The Middle Ages of the internal-combustion engine 1794 – 1886.* loc cit

Hewison, C. H. 1983. *Locomotive boiler explosions*. Newton Abbot, Devon: David & Charles

Heyman, J. 1980. The estimation of the strength of masonry arches. *Proc. Institution of Civil Engineers*, 69(2):921-37, Dec

Hibbs, J. 1968. *The history of British bus services*. Newton Abbot, Devon: David and Charles

Hill, P. 1980. The League's earliest years. *American Wheelmen*. Centennial issue, 16(5):8-14, May

Hilton, G. W. 1969. Transport technology and the urban pattern. *J Contemporary History*, 4:123-35

Hirst, J. 2009. *The shortest history of Europe*. Collingwood, Victoria: Black

Hodgkinson, E. 1840. Experimental researches on pillars of cast iron, and other materials. *Philosophical Trans Roy Soc*, 130, Paper XX, Jan, 385-454

Hopkins, H. J. 1970. *A span of bridges*, Newton Abbot, Devon: David & Charles

Humber, W. 1857. *A practical treatise on cast and wrought iron bridges and girders*. London: Spon

Hussey, A. 2006. *Paris: the secret history*. London: Viking

Inwood, S. 2005. *City of cities: the birth of modern London*. London: McMillan

Jackson, G. 1981. The ports. In Aldcroft and Freeman, 177-209

Kelly, A. & M. (Ed) 2006. *Brunel: in love with the impossible*. Bristol: Bristol Cultural Development Council

Kitson Clark, G. 1962. *The making of Victorian England*. London: Methuen

Kurrer, K-E. 2008. *The history of the theory of structures*. Berlin: Ernst & Son

Kuzmanovic, B. 1977. History of the theory of bridge structures. *Proc American Society of Civil Engineers* 103(ST5):1095-1111

Kuznets, S. 1930. *Secular movements in production and prices*. Boston and New York: Houghton Mifilin

Labrum, E. A. 1994. Eastern & central England. *Civil engineering heritage*. London: Thomas Telford

Lamm, R., Psarianos, B. & Mailaender, T. 1999. *Highway design and traffic safety engineering handbook*. New York: McGraw Hill

Lane, R. H. 1935. Waggons and their ancestors. *Antiquity*, 9:140-150

Law, H. 1850. *Rudiments of the art of constructing and repairing common roads*. London: John Weale Includes S, Hughes. A general survey of the principal metropolitan roads.
Lay, M. G. 1982a. *Structural steel fundamentals*. Vermont Sth (Melbourne): Australian Road Research Board
—. 1982b. *Source book for the Australian steel structures code AS 1250*. Sydney: Australian Institute of Steel Construction.
—. 1988. The report literature. *EI Insider* (Engineering Index), 2nd quarter 10:5-6
—. 1991. Design support, or Degin Opprsstu. *The Art of Structural Engineering, University of Melbourne Conference in honour of L. K. Stevens*:59-68
—. 1992. *Ways of the world*. New Brunswick: Rutgers University Press
—. 1998. The history of vehicle suspensions. Routes/Roads 298-II: 56-63, April
—. 2003a. *Melbourne Miles*. Melbourne: Australian Scholarly Press
—. 2003b. *That dinkum breed of true Aussies*. Melbourne: the author (ISBN 0-9751943-0-5)
—. 2004. History of traffic signs. Ch 2. *The human factors of traffic signs*. Boca Raton, Fl: CRC press (Eds C. Castro & T. Horberry)
—. 2005.The history of transportation planning. in *Handbook of transport strategy, policy and institutions*, p157-173, Vol 6 in series, Oxford: Elsevier. Button, K. & Hensher, D. (Eds).
—. 2007. Review of the major factors influencing road development during the XXth century. 2007 PIARC World Congress Historical Symposium SP18/19. Paris
—. 2008, A history of freight transport prior to the modern truck, *Proc International Conf on Heavy Vehicles*, Paris (B. Jacob, et al editors), ISTE & Wiley, London & New York, pp 35-47
—. 2009. *Handbook of road technology*. London: Spon (Taylor & Francis)
—. 2010. *Strange ways*. Sydney: EA Books:p68
—. 2011. Measuring traffic congestion. *Road & Trsp Res*, 20(2):50-61, June
—. 2012a. An historical view of the assessment and management of congestion. *Road & Trsp Res*, 21(2):32-45, June
—. 2012b. How were motorways specified? A comment on the special section on roads. *Transfers: interdisciplinary journal of mobility studies*, 2(1):134-140, Spring
Lessing, H-E, 2003. *Automobilität: Karl Drais und die unglaublichen Anfänge*. Leipzig: Maxime, verlag Maxi Kutschera

Lewis, M. J. 1970. *Early wooden railways*. London: Routledge Kegan Paul
Lucier, P. 1991. Petroleum: what is it good for? *Invention & Technology*: 7(2):56-63
Maceroni, F. 1835. *Expositions and illustrations interesting to all those concerned in steam power*. London: Effingham Wilson (available as a Google Play ebook)
Martin, T. & Macleod, I. A. 1995. The Tay rail bridge disaster – a reappraisal based on modern analysis methods. *Proc Inst Civ Engrgs, Civ Engrg*, May, 108:77-83
Marx, K. 1857. *The Grundrisse: Outlines of the Critique of Political Economy*. Notebook V:538 (Penguin Classics 1993)
McAdam, J. L. 1816+. *Remarks (or Observations) on the present system of roadmaking*. 1st edition (9th edition 1827). London: Longman, Rees, Orme, Brown and Green.
McCausland, H. 1948. *The English carriage*. London: Batchworth
McShane, C. 1994. *Down the asphalt path – the automobile and the American city*. New York: Columbia UP
Mercedes-Benz AG 1992. *Mercedes-Benz Museum*. Stuttgart: Merces-Benz AG
Meldahl, K. H. 2007. *Hard road west*. Chicago: Univ of Chicago Press
Melosi, M. V. 1982. *Energy transitions in the nineteenth century economy*. In Daniels & Rose 1982:55-69
Mom, G. 2007. *Constructing multifunctional networks; road building in the Netherlands, 1810-1980*. In Road history: planning, building and use. Neuchâtel; Alphil Eds Mom & L. Tissot)
—. 2012. Editorial. *Transfers* 2(1):1-4
Moore, H. C. 1902. *Omnibuses and cabs: their origin and history*. London: Chapman and Hall
More, C. 2000. *Understanding the Industrial Revolution*. London: Routledge
Mountfield, D. 1976. *The coaching age*. London: Robert Hale
Mumford, L. 1934. *Technics and civilisation*. New York: Harcourt Brace
Nasmyth, J. 1897. *John Nasmyth: Engineer, an autobiography*. (S. Smiles, Ed). London: John Murray
Navier, C. L. 1823. *Memoire sur les ponts suspendu*. Paris, Imprimerie Royale
Neuberg, V. E. 1977. *Popular literature: history and culture*. Harmondsworth, Middlesex: Penguin
Niemann, H. 1994. *Benz & Cie*. Stuttgart: Motorbuch

Nock, O. S. 1972. *The dawn of world railways: 1800-1850*. London: Blandford

Pannell, J. P. M. 1964. *An illustrated history of civil engineering*. London: Thames & Hudson

Parnell, H (Sir) 1838. *A treatise on roads*. London: Longman, Orme, Brown, Green and Longmans

Parsons, N. T. 2007. *Worth the detour – a history of the guidebook*. Gloucestershire: Sutton

Paxton, R. & Shipway, J. 2007a. *Scottish lowlands and borders - Civil engineering heritage*. London: Thomas Telford

—. 2007b. *Scottish highlands and islands- Civil engineering heritage*. London: Thomas Telford

Penfold, A., editor, 1980. *Thomas Telford: Engineer*. London: Thomas Telford.

Perkin, H. 1969. *The origins of modern English society: 1780-1880*. London: Routledge & Kegan Paul

Peters, T. F. 1996. *Building the nineteenth century*. Cambridge (Mass): MIT Press

Petroski, H. 1994. Success syndrome: the collapse of the Dee Bridge. *Civ Engrgr* (ASCE), 64(4): 52-5

—. 1996. Harnessing steam. *Am Sci* 84(1):15-9

—. 2007. Why things break. *Am Sci* 95(3): 206-9

Pettifer, J. & Turner, N. 1984. *Automania*. London: Collins

Picon, A. 2004. Engineers and engineering history: problems and perspectives. *J History & Technology* 20(4): 421-436

Plowden, W. 1971. *The motor car and politics 1896-1970*. London: Bodley Head

Plumb, J. H. 1950. *England in the 18th century*. Harmondsworth, Middlesex: Penguin

Pritchard, W.G. & J.K. 1994. Mathematical models of running. *American Scientist* 82(6):546-53, Nov-Dec

Pyne, W. H. 1806. *Microcosm*: or Picturesque groups for the embellishment of landscape, drawn and etched by Pyne, tinted by J. Hill and with words by C. Gray. 2 vols. London: Nattes. There were many subsequent editions. The facsimile edition by Blom in New York in 1971 is based on a collected version published by Nattali of London in 1845.

Raminez, A. 2016. Bessemer's volcano and the birth of steel. *American Scientist* 104(1):12-5, Jan-Feb

Rankine, W. J. M. 1843. On the causes of unexpected breakage of the journals of railway axles. Minutes of the *Proc Inst Civ Engrgs* II:105-8

—. 1908. *Useful rules and tables*. London: Griffin

Ransom, P. J. 1984. *The archaeology of the transport revolution*, 1750-1850. Kingswood, Surrey: World's Work

—. 1990. *The Victorian railway and how it evolved*. London: Heinemann

Raymond, R. 1984. *Out of the fiery furnace: the impact of metals on the history of mankind*. Melbourne: Macmillan

Reader, W. J. 1980. *McAdam---the McAdam family and the turnpike roads:1798-1861*. London: Heinemann.

Reed, M. 2004. Hudson, George [the Railway King] (1800–1871). *Oxford Dictionary of National Biography*, Oxford: Oxford University Press

Rogers, H. C. 1961. *Turnpike to iron road*. London: Seeley, Service and Co.

Rolt, L. C. 1958. *Thomas Telford*. London: Longmans

Rosen, W. 2010. *The most powerful idea in the world*. New York: Random House

Rostow, W. W. 1990. *The stages of economic growth*. Cambridge: Cambridge University Press

Ruiz, M. 1988. *The history of the automobile*. New York: Gallery. *First published in 1984 in Italian*.

Ryall, M. J. 1999. Britannia Bridge: from concept to construction. *Proc Inst Civ Engrs* 132:132-143, May

Sachs, W. 1992. *For love of the automobile*. Berkeley: University of California Press. *First published in 1984 in Germany.*

Schivelbusch, W. 1979 (German Edn 1977). *The railway journey – the industrialisation of time and space in the 19th century*. Berkeley: Urizen & University of California Press

Schreiber, H. 1961. *The history of roads*. London: Barrie & Rockliff (Stewart Thomson, translator) (originally published in German by Econ-Velag in Düsseldorf, Germany in 1961 as *Sinfonie der Strasse*)

Schumpeter, J. A. 1911. *The theory of economic development*. London (U.K.): Transaction Publishers.

Searight, T. B 1894. *The Old Pike: a history of the national Road*. Uniontown, Pa: The Author (1990 reprint by Heritage Books, Bowie, Md)

Sekon, G. A. 1938. *Locomotion in Victorian London*. London: Oxford University Press

Simmons, J. 1975. *Rail 150:the Stockton and Darlington Railway and what followed*. London: Eyre Methuen. Simmons is also editor; other contributors are Hoole, K. & Bonavia, M.

Simsa, P. 1986. Mobility. *In aller Welt (Mercedes-Benz)*. 199(1):4-21

Singer, C., Holmyard, E., Hall, A. & Williams, T. 1958. *A history of technology, Volume 5, The late 19th century, c1850 to c1900*. Oxford: Oxford University Press

Smil, V. 2005. *Creating the twentieth century: technical innovations of 1867-1914 and their lasting impact*. Oxford: Oxford University Press

—. 2017. *Energy and civilisation: a history*. Cambridge (Mass): MIT Press

Smiles, S. 1867. *Lives of the engineers. The life of Thomas Telford*. London: John Murray

—. 1874. *Lives of the engineers. History of roads. Metcalfe: Telford*. London: John Murray

Smith, D. 2001. *London and the Thames valley. Civil engineering heritage*. London: Thomas Telford

Sobel, D. 1996. *Longitude*. London: Fourth Estate (later Harper Perennial)

Spencer, H. 1896. *The Study of Sociology*, New York: D Appleton

Steinman, D. B. & Watson, S. R. 1957. *Bridges*. 2nd Edn. New York: Dover

Straub, H. 1952. *A history of civil engineering*. London: Leonard Hill

Syme, R. 1952. *The story of Britain's highways*. London: Pitman

Tarr, L. 1969. *The history of the carriage*. London: Vision (translated by E. Hoch)

Telford, T. 1838. *Thomas Telford, civil engineer, his life by himself*. London: J. and L.G. Hansard and Sons

Thompson, E. P. 1963. *The making of the English working class*. London: Victor Gollancz

Thompson, F. M. L. 1963. *English landed society in the nineteenth century*. London: Routledge Kegan Paul

Thompson, J. & Smith, A. 1877. *Street Life in London*, London: Sampson Low, Marston, Searle and Rivington

Timoshenko, S. 1953. *History of strength of materials*. New York; McGraw Hill

Tredgold, T. 1924. *Practical essay on the strength of cast iron, and other metals*. 2nd Edn. London: J Taylor (the first edition was in 1822 and published by Weales).

Trinder, B. 1979. *The Iron Bridge*. Coalbrookdale: Iron Bridge Gorge Museum Trust. Museum Guide 3.01

Tucker, V. A. 1975. The energetic cost of moving about. *Am Sci* 63(4):413-9, July

Uglow, J. 2002. *The Lunar Society: five friends whose curiosity changed the world*. London: Faber & Faber

Vale, E. 1960. *The mail-coach men of the late eighteenth century*. London: Cassell

Van der Merwe, N. J. & Avery, D. H. 1982. Pathways to steel. *Am Sci* 70(2):146-155

Vitruvius, M. (c50BC). *De Architectura*. M. H. Morgan's translation. 1960, New York: Dover

Watkins, J. E. 1891. The beginnings of engineering. *Trans ASCE* XXIV(472), May, 78pp

Watney, M. 1961. *The elegant carriage*. London: J A Allen

Whitfield, P. 2006. *London: a life in maps*. London: British Library

Williams, F. S. 1885. *Our iron roads: their history, construction and administration*. (6th edn, 1st edn in 1835). London: Bemrose

Wise, S. 2008. *The blackest streets*. London: Bodley Head

Witcher, T. R. 2017. The birth of steam railroads: the Charleston-Hamburg railroad. *ASCE Civil Engrg* 87(9):44-7, Sept

Wosc, J. 1992. *Breaking frame: technology and the visual arts in the nineteenth century*. New Brunswick (NJ): Rutgers Uni Press

Young, A. 1771. *Six months tour through the north of England*. Edinburgh: W. Strahan et al

—. 1785. *Annals of agriculture and other useful arts*. Vol 4. London: Bury St Edmund

Youngson, A. J. 1967. *Overhead capital: a study of development economics*. Edinburgh: Edinburgh University Press

INDEX